STUDY GUIDE
TO ACCOMPANY

ENVIRONMENT

third edition

Peter H. Raven

Missouri Botanical Garden

Linda R. Berg

St. Petersburg Junior College

Elizabeth E. Reeder

Loyola College in Maryland

HARCOURT COLLEGE PUBLISHERS

Fort Worth   Philadelphia   San Diego   New York   Orlando   Austin   San Antonio
Toronto   Montreal   London   Sydney   Tokyo

Printed in the United States of America

ISBN 0-03-031579-4

012 202 7654321

Preface

The intention of this study guide is to serve as an aid to learning and as a review that follows careful reading of each chapter. The text will be needed for its prose and figures as you make your way through the study guide, so keep the big book handy.

Chapter content is covered by multiple choice, matching, fill-in, short-answer, critical thinking, and data-analysis (quantitative) questions. Multiple-choice, matching, and fill-in questions focus primarily on vocabulary and factual information, in other words, the background knowledge that serves as a foundation for the chapter's primary message.

Short-answer questions demand integration and synthesis of information; please don't approach them with memorization in mind. If you can *speak* the response, as if explaining something to a friend, you understand it. And, remember, the answers given are subject to improvement; your own approach may be better and will certainly be easier to recall.

Critical-thinking questions are open-ended; they do *not* have "right" and "wrong" answers. Therefore, the answer section provides only hints, in the form of questions, intended to serve as springboards. Critical-thinking questions are meant to stimulate your thinking, tweak your imagination, and probe your values. If you study with friends, they may even generate some lively disagreements.

Data-analysis questions are based on quantitative information given in the text. Even if you don't like math, the world we have constructed for ourselves requires some basic mathematical skills to negotiate. The time you spend on these questions is likely to help you in other subjects as well, and improve your ability to find meaning in the numerical data forever popping up in articles, political debates, advertisements, and a multitude of other sources.

Please use this study guide to your advantage; it is intended to be your ally. The explosion of information is a reality of our times, and we all need all the help we can get to master as much of it as quickly and efficiently as possible. More importantly, this text, *Environment*, raises some of the most urgent and challenging issues our species and our planet have ever faced. Everyone needs to know not only what is happening but what needs to be done to halt environmental destruction and begin to mend the damage. So, thank you for your willingness to be among the "informed." Earth's future depends on you. Study well.

Contents

PART FIVE: OUR PRECIOUS RESOURCES

PART SIX: ENVIRONMENTAL CONCERNS

PART SEVEN: TOMORROW'S WORLD

PART ONE

HUMANS IN THE ENVIRONMENT

Chapter 1

Our Changing Environment

Study Outline:

I. Environmental Science
    A. Environmental Sustainability

II. Our Impact on the Environment
    A. Increasing Human Numbers
    B. Endocrine Disrupters
    C. Closing of the Georges Bank Fishery
    D. Declining Bird Populations
    E. Reintroducing Wolves to Yellowstone
    F. The Introduction of Exotic Species
    G. Damage to the Atmosphere: Stratospheric Ozone Depletion
    H. Global Climate Warming and Increasing Carbon Dioxide Levels
    I. Destroying a Tropical Rain Forest

III. The Goals of Environmental Science

IV. Meeting the Challenge: The Earth Summit

V. Envirobrief: Coffee That's For the Birds

VI. Envirobrief: Welcome Home

VII. Summary with Selected Key Terms

Key Terms: The following terms are listed in order of appearance in your textbook.

green architecture
environmental science
pollution
environmental sustainability

poverty
endocrine disrupters
hormones
synergism
antagonism
Food Quality Protection Act
Safe Drinking Water Act
commercially extinct
Magnuson Fishery Conservation and Management Act
subsidy
tropical migrant songbirds
forest edge
nest parasitism
Endangered Species Act (ESA)
stratosphere
chlorofluorocarbons (CFCs)
Kyoto Protocol
watersheds
bio-prospect
ecotourism
ecology
deforestation
sustainable development

Multiple Choice:

1. The ability of the environment to function indefinitely without declining from pressures imposed by humans is
    a. environmental sustainability
    b. environmental science
    c. synergism
    d. antagonism
    e. ecosystem management

2. Of the nearly 400 cities with populations of 1 million or more, 284 are in
    a. highly developed countries
    b. developing countries
    c. the tropics
    d. the Western Hemisphere
    e. Asia

3. In 1999, the world's human population reached (and passed)
   a. 300 million
   b. 600 million
   c. 1 billion
   d. 3 billion
   e. 6 billion

4. Based on present trends, world population is expected to _____ by the end of the 21st century.
   a. exceed 25 billion
   b. stabilize between 10.4 billion and 20 billion
   c. stabilize between 6 billion and 9 billion
   d. decline by 2 billion
   e. decline by 4 billion

5. Pollutants called endocrine disrupters mimic _____ in humans and other animals.
   a. proteins
   b. DNA
   c. hormones
   d. pheromones
   e. phospholipids

6. Endocrine disrupters are often associated with _____ disorders.
   a. nervous system
   b. digestive
   c. reproductive
   d. behavioral
   e. respiratory

7. Two or more pollutants interacting in such a way that their combined effects are greater than the sum of their individual effects illustrate
   a. hormone-mimicking effects
   b. synergism
   c. antagonism
   d. biomagnification
   e. bioaccumulation

8. The collapse of fish populations, which led to the closing of the Georges Bank fishing ground, was primarily the result of
   a. fish diseases
   b. water pollution
   c. diminished plankton
   d. overfishing
   e. over-regulation

9. Reauthorization of the Magnuson Fishery Conservation and Management Act in 1997 included passage of a new law requiring strategies to help declining _____ populations recover.
   a. shellfish
   b. sea turtle
   c. phytoplankton
   d. marine fish
   e. marine mammal

10. Government _____ help(s) make the fishing industry profitable.
    a. regulations
    b. subsidies
    c. supervision
    d. non-interference
    e. taxes on fishermen

11. Tropical migrant songbirds are
    a. in immediate danger of extinction
    b. threatened by habitat loss
    c. birds of prey
    d. dwellers of forest edges
    e. nest parasites

12. Nest predation and nest parasitism increase
    a. near water
    b. in large, unbroken forest tracts
    c. near forest edges
    d. as day length increases
    e. as temperature increases

13. Cowbirds harm songbirds by
    a. eating their eggs
    b. spreading diseases
    c. competing for food sources
    d. competing for nesting sites
    e. laying their eggs in songbirds' nests

14. In 1995, wolves were reintroduced to parts of their historic ranges in _____ and _____.
    a. Wyoming, Idaho
    b. Washington, Oregon
    c. California, Nevada
    d. Montana, Nebraska
    e. Minnesota, Wisconsin

15. The Defenders of Wildlife has a "wolf compensation fund" that is used to
    a. pay for wolf reintroduction expenses
    b. pay for the costs of legal battles involving reintroduced wolves
    c. reimburse ranchers who lose livestock to wolves
    d. educate the public about the value of wolves to ecosystems
    e. replace elk killed by wolves

16. Ballast water released from ships often
    a. contains toxic wastes
    b. contains pathogens (disease-causing organisms)
    c. is polluted with diesel fuel
    d. spreads exotic species
    e. floods wetlands

17. A _____ from North America has caused devastating declines in fish populations of the Black Sea.
    a. mussel
    b. comb jelly
    c. water mold
    d. fish virus
    e. coral

18. The zebra mussel is a(n) _____ species found in the Great Lakes and many other aquatic habitats in North America.
    a. endangered
    b. exotic
    c. native
    d. threatened
    e. commercially extinct

19. Chlorofluorocarbons (CFCs)
    a. absorb UV light
    b. occur naturally in the upper atmosphere
    c. cause thinning of the ozone layer
    d. are highly toxic
    e. come from car exhaust

20. Skin cancer is one of the possible results of exposure to
    a. ozone
    b. endocrine disrupters
    c. carbon dioxide
    d. ultraviolet radiation
    e. CFCs

21. The burning of forests and fossil fuels adds _____ to the atmosphere, which contributes to global _____.
    a. carbon dioxide, warming
    b. carbon dioxide, cooling
    c. ozone, warming
    d. ozone, cooling
    e. none of the above

22. The warmest year recorded since 1860, when global weather data began to be systematically recorded, was in the
    a. mid 1980s
    b. late 1980s
    c. early 1990s
    d. mid 1990s
    e. late 1990s

23. The Kyoto Protocol is an international treaty that addresses reduction of
    a. the global extinction rate
    b. carbon dioxide emissions
    c. CFC use
    d. human numbers
    e. competition for water

24. Tropical deforestation makes it harder for migratory songbirds to
    a. build up the food reserves needed for migration
    b. establish adequate territories
    c. breed successfully in North America
    d. maintain their numbers
    e. all of the above

25. Forest trees provide long-term storage of
    a. water
    b. oxygen
    c. carbon dioxide
    d. ozone
    e. sulfur

26. Environmental science is primarily a(n) _____ science, with a focus on _____.
    a. applied, problem solving
    b. theoretical, establishing general principles
    c. applied, establishing general principles
    d. theoretical, solving problems
    e. applied, understanding species interactions

27. *Agenda 21*, first discussed at the 1992 Earth Summit in Rio de Janeiro, is a complex plan with the primary goal of
    a. sustainable development
    b. a lowered extinction rate
    c. better human health
    d. reforestation
    e. environmental education

28. Worldwide, an area the size of _____ is deforested every year.
    a. Long Island
    b. Kentucky
    c. Michigan
    d. Texas
    e. Alaska

Matching: Match the terms on the left with the responses on the right.

_____1. fossil fuels
_____2. endocrine disrupters
_____3. hormones
_____4. estrogens
_____5. haddock
_____6. cerulean warbler
_____7. opossum
_____8. cowbird
_____9. zebra mussel
_____10. Kyoto Protocol
_____11. shade-grown coffee
_____12. Earth Charter

a. has reached commercial extinction in Georges Bank area
b. a nest parasite
c. chemical messengers that regulate many biological functions
d. an exotic species introduced to the Great Lakes
e. nonrenewable resources
f. an international treaty with the goal of decreasing carbon dioxide emissions
g. a migratory songbird species in decline
h. include certain plastics, chlorine compounds, heavy metals, and pesticides
i. a philosophical statement about environment and development; the "Rio Declaration"
j. female sex hormones
k. like the cowbird, an "edge species"
l. benefits migratory songbirds

Fill-In:

1. _____ (two words) is a growing trend that encompasses energy and water conservation, recycled building materials, and other environmentally friendly aspects of building design.

2. _____ (two words) is the interdisciplinary study of our relationship with other species and the physical environment.

3. The world's human population is expected to increase by _____ billion in the next 30 to 40 years.

4. A human population's size and level of _____ largely determine its impact on the environment.

5. Studies of human sperm counts between 1940 and 1990 showed a more than _____ % decline.

6. Soy-based foods are examples of foods containing natural _____.

7. If the combined effects of two interacting pollutants are less severe than the sum of their individual effects, they illustrate a phenomenon called _____.

8. The 1996 amendments to the Food Quality Protection Act and the Safe Drinking Water Act require that the U.S. Environmental Protection Agency develop a plan to test chemicals for their potential to disrupt the _____ system.

9. When a species or population reaches numbers too low for profitable harvest, it is said to be _____ (two words).

10. Wood thrushes and yellow-billed cuckoos are examples of _____ (two words) songbirds.

11. _____ of North American forests diminishes the reproductive success of many migratory songbirds.

12. _____ (two words) cluster on piers, buoys, the hulls of boats, and water-intake systems; they also deplete the food sources of other species.

13. The North American Free Trade Agreement increases the risk of invasion by _____ species.

14. Ozone in the _____ (layer of the atmosphere) is destroyed by _____.

15. Ozone absorbs _____ radiation.

16. Recovery of the ozone layer is not expected until the middle of the _____ century.

17. _____ (two words) in the atmosphere traps heat that would normally radiate into space.

18. The U.S. National Academy of Science predicts that global _____ (two words) will be the most pressing international problem of the 21st century.

19. All forests are _____ (areas drained by river systems); as such, they protect water quality.

20. _____ is the discipline of biology that studies interrelationships between organisms and their environment.

21. Yellowstone wolves have decimated some populations of _____.

Short Answer:

1. What are some of the ways human society is not operating in an environmentally sustainable manner?

2. Besides cod, haddock, and yellowtail flounder of Georges Bank, name some fisheries that have declined in recent years or recent decades.

3. Why does overfishing occur?

4. Name several examples of tropical migrant songbird species in decline.

5. What were the two main goals of the wolf-reintroduction effort at Yellowstone National Park?

6. How did the introduction of comb jellies to the Black Sea harm native species and the local economy?

7. What are some of the predicted results of global warming?

8. Why do critics oppose the Kyoto Protocol, and how do energy experts respond to such criticisms?

9. Name some of the forest organisms that died or suffered diminished likelihood of survival as a result of the Brazilian fire depicted in Figure 1-11 in *Environment*, 3/e.

10. What does a company do when it *bio-prospects* in a forest?

11. What global environmental issues were addressed at the 1992 U.N. Conference on Environment and Development (Earth Summit)?

Critical-Thinking Questions:

1. Make a list of all the animals and plants you have relied on today. Are they all domesticated species? If not, which are not?

2. For what reason was the Yellowstone gray wolf declared an "experimental nonessential" species? Do you agree or disagree with the reasoning behind this designation?

3. Explain how the reintroduction of wolves *could* lead to an increased number of pond and wetland habitats in Yellowstone National Park.

Data Interpretation:

1. The World Bank estimates that the number of people (out of a total population of approximately 6 billion) living in extreme poverty is 1.3 billion. What percentage of the total human population is living in extreme poverty?

2. The world's human population was about 1 billion in 1800; it reached 6 billion in 1999 (see Figure 1-2 in *Environment*, 3/e). How many times did the world population double between 1800 and 1999?

3. In Florida's Lake Apopka, alligators have very high egg and infant mortality because of a chemical spill in 1980. Only 40% of eggs hatch, and only 50% of hatchlings survive beyond 10 days. Suppose 3000 eggs are laid. How many surviving hatchlings would be expected on the 11th day after hatching?

Chapter 2

Addressing Environmental Problems, Part I

Study Outline:

I. Addressing Environmental Problems: An Overview

II. The Scientific Analysis of Environmental Problems
   A. The Scientific Method
   B. Inductive and Deductive Reasoning
   C. The Importance of Prediction
   D. Experimental Controls
   E. Theories and Principles

III. Scientific Decision-Making and Uncertainty: An Assessment of Risks
   A. Determining the Health Effects of Environmental Pollutants

IV. Ecological Risk Assessment
   A. Cost-Benefit Analysis of Risks
   B. A Balanced Perspective on Risks

V. Case In Point: The Lake Washington Story
   A. Birth of an Environmental Problem
   B. Sounding the Alarm
   C. Scientific Assessment
   D. Risk Analysis
   E. Public Education
   F. Political Action
   G. Follow-Through

VI. Working Together

VII. Envirobrief: Environmental Literacy

VIII. Envirobrief: No Quick Fix for the Salton Sea

IX. Focus On: The Tragedy of the Commons

X.   Summary with Selected Key Terms

Key Terms: The following terms are listed in order of appearance in your textbook.

model
data
scientific method
hypothesis
inductive reasoning
deductive reasoning
variable
control
theory
principles
law
risk
risk assessment
risk management
toxicology
dose
response
lethal dose-50% (LD_{50})
effective dose-50% (ED_{50})
dose-response curve
carcinogen
ecological risk assessments
stressors
cost-benefit analysis
eutrophication
global commons
stewardship

Multiple Choice:

1.   _____, which infects hundreds of millions of people each year and causes several
     million deaths, is expanding its geographical range because of environmental
     changes.
     a. Cholera
     b. Tuberculosis
     c. Sleeping sickness
     d. Typhus
     e. Malaria

2. The scientific process _____ establish(es) proof; it _____ claim(s) to know "final answers."
   a. can, can
   b. cannot, cannot
   c. can, cannot
   d. cannot, can
   e. always, frequently

3. After studying several different sites and species, a researcher concludes that certain songbird populations decline in areas with high cowbird populations. This conclusion has been arrived at through use of
   a. inductive reasoning
   b. deductive reasoning
   c. hypothetical conjecture
   d. risk assessment
   e. cost-benefit analysis

4. In preparing to test a hypothesis, a researcher uses _____ to determine the type of experiment or observations that will be most useful.
   a. inductive reasoning
   b. deductive reasoning
   c. hypothetical conjecture
   d. risk assessment
   e. cost-benefit analysis

5. The researcher in question #3 compares songbird declines in areas with cowbirds with songbird population trends in areas *without* cowbirds. Collecting data from areas without cowbirds serves as the experimental
   a. independent variable
   b. dependent variable
   c. theory
   d. control test
   e. hypothesis

6. When speaking of the "theory of evolution" or the "theory of relativity" (or any other scientific theory), we speak of an explanation that is
   a. highly conjectural
   b. supported by only a few observations
   c. supported by only a few experiments
   d. supported by a large body of observations and experiments
   e. proven as absolute fact

7. A risk probability of .25 is a _____ risk than a risk of .30; a risk probability of 1 means _____ risk.
   a. higher, certain
   b. lower, certain
   c. higher, no
   d. lower, no
   e. higher, uncertain

8. In general, we seem to be most afraid of risks
   a. that pose the highest likelihood of harm
   b. over which we have little or no control
   c. we cannot measure accurately
   d. that are newly introduced to our lives
   e. to species other than our own

9. Someone who develops frequent cases of bronchitis as a result of breathing air badly polluted with sulfur dioxide is experiencing a _____ dose of sulfur dioxide with _____ toxicity.
   a. lethal, acute
   b. lethal, chronic
   c. sublethal, acute
   d. sublethal, chronic
   e. sublethal, no

10. The only diseases traditionally evaluated in chemical risk assessment have been
    a. kidney diseases
    b. liver diseases
    c. infectious diseases
    d. food-poisoning diseases
    e. cancers

11. The link between cancer and dioxins was first recognized through
    a. laboratory tests on animals
    b. controlled experiments using human subjects
    c. epidemiological evidence
    d. observational experiments involving chemical mixtures
    e. a proliferation of lawsuits stemming from a chemical spill

12. If mixing a chemical with a toxicity level of 2.0 with a chemical with a toxicity level of 1.5 gives a resulting mixture with a toxicity level of 3.5, the chemical mixture is
    a. synergistic
    b. antagonistic
    c. additive
    d. nontoxic
    e. sublethal

13. Examination of a stressor-response profile is part of the _____ phase of ecological risk assessment.
   a. problem formulation
   b. analysis
   c. risk characterization
   d. risk management
   e. mitigation

14. Recently, the EPA has used _____ to examine the effects of stressors on species in the Snake River ecosystem.
   a. ecological risk assessment
   b. cost-benefit analysis
   c. mitigation
   d. risk management
   e. all of the above

15. From the 1920s until the 1960s, Washington State's Lake Washington suffered declining water quality, primarily as the result of
   a. industrial wastes
   b. sewage effluent
   c. repeated oil spills
   d. urban runoff
   e. farm runoff

16. Overabundance of a type of _____ in Lake Washington was one of the warning signs of its decline.
   a. green alga
   b. filamentous cyanobacterium
   c. eel
   d. jellyfish
   e. parasitic protozoan

17. The cause for the increase in the species mentioned in #16 was recognized to be
   a. a decline of predators
   b. heavy-metal pollution
   c. cloudy weather
   d. dissolved nutrients
   e. lack of dissolved oxygen

18. Which of the following is a typical symptom of eutrophication?
   a. green scum on the water's surface
   b. reduced oxygen levels
   c. a bad smell
   d. declining fish and invertebrate populations
   e. all of the above

19. In the 1950s, W. T. Edmondson of the University of Washington pointed out that *Ocsillatoria rubescens* is a species associated with
    a. pristine water
    b. a recovering aquatic ecosystem
    c. polluted water
    d. overfishing
    e. diseased fish populations

20. In the 1960s, the solution to Lake Washington's deterioration was to
    a. halt development in its watershed
    b. build state-of-the-art wastewater treatment facilities
    c. transport treated sewage to Puget Sound
    d. fund further research
    e. prosecute polluting industries

21. To pay for treatment of sewage and transport of effluent in the Lake Washington area,
    a. local citizens sponsored fairs and other fund raisers
    b. state legislators requested federal money
    c. local businesses and industries were required to pay additional taxes
    d. local householders were required to pay additional taxes
    e. fees for fishing licenses were increased

22. With the decline of *Oscillatoria*, Lake Washington's water became clearer (more transparent) than expected because *Daphnia*, which _____ nonfilamentous algae, _____ in number.
    a. eat, decreased
    b. eat, increased
    c. are eaten by, decreased
    d. are eaten by, increased
    e. compete with, increased

23. In his famous essay published in 1968, Garrett Hardin used _____ in medieval Europe to illustrate the tragedy-of-the-commons effect.
    a. overgrazing of shared pastureland
    b. human overpopulation
    c. overhunting
    d. the spread of infectious disease
    e. crop failures

24. California's Salton Sea is an important habitat for _____; however, hundreds of them have died from _____ believed to be related to the water's chemistry.
    a. freshwater fishes, parasitic infections
    b. freshwater fishes, gill damage
    c. amphibians, skin lesions
    d. birds, diseases
    e. birds, lack of dietary sodium

25. In this figure (Figure 2-7b in *Environment*, 3/e), phosphorus represents amount of _____, and chlorophyll represents _____.
    a. sewage effluent, number of cyanobacteria
    b. sewage effluent, dissolved-oxygen level
    c. algae, number of cyanobacteria
    d. algae, dissolved-oxygen level
    e. detergent, amount of sewage effluent

Matching: Match the terms on the left with the responses on the right.

____1. model	a. an integrated explanation of numerous hypotheses
____2. risk analysis	b. the type and amount of damage caused by exposure to a particular dose
____3. data	c. deals with the potential effects of intervention
____4. hypothesis	d. part of experimental design that holds constant the variable being tested
____5. variable	e. a formal statement that describes a situation; constructed via scientific assessment
____6. control test	f. based on well-established theories
____7. theory	g. the process of nutrient enrichment of freshwater lakes
____8. principles	h. objective information collected by observation and experimentation
____9. risk	i. the probability of harm
____10. dose	j. a cancer-causing substance
____11. response	k. an educated guess that explains a problem or phenomenon
____12. carcinogen	l. human-induced changes that tax the environment
____13. human-induced stressors	m. a factor that influences a process
____14. eutrophication	n. the amount of a toxin that enters the body of an exposed organism

Fill-In:

1. The first stage of addressing any environmental problem is _____ (two words).

2. Although most people view science as a body of knowledge, science is also a _____.

3. _____ (two words) is the established process used by scientists to answer questions and solve problems.

4. _____ reasoning involves discovering general principles by examining specific cases or examples.

5. _____ reasoning proceeds from general principles to specific cases.

6. Experimental science must meet the challenge of designing _____ tests and isolating a single _____ for study.

7. _____ (two words) involves quantifying risks of a certain action so they can be compared and contrasted with other risks.

8. _____ (two words) includes the development and implementation of laws to regulate hazardous substances.

9. _____ is the study of chemicals with adverse effects on health.

10. A chemical, such as cocaine, with a low lethal dose-50% is _____ (more or less) toxic than a chemical, such as morphine, with a higher lethal dose-50%.

11. The dose that kills half of a population of test animals is called the (two words) _____-50%.

12. The dose that elicits a particular response in half of a population is called the (two words) _____-50%.

13. A _____ curve shows the effect of different doses on a population of test organisms.

14. Cigarette smoke and car exhaust are examples of chemical _____, which are difficult for toxicologists to evaluate.

15. A(n) _____ chemical mixture has a greater combined effect than expected; a(n) _____ chemical mixture has a smaller combined effect than expected.

16. Doing risk assessments for the environment is _____ (more or less) difficult than doing risk assessments that involve human health.

17. Risk assessment that looks at environmental, as opposed to human-health, effects is called _____ risk assessment.

18. _____ analysis involves comparison of estimated costs with potential benefits to determine how much of a particular toxin or pollutant is tolerable.

19. The decline of Lake Washington's filamentous _____ and nonfilamentous _____, due to reduced water pollution and an increase in _____ populations, greatly improved water clarity.

20. Individuals, businesses, and governments must foster a sense of _____, or shared responsibility for the sustainable care of the planet.

21. The LD_{50} (in mg/kg) for caffeine is 200.0; for nicotine, it is 53.0. Which substance is more toxic? _____

Short Answer:

1. Explain how environmental change increases the incidence of malaria.

2. What are the five basic components of environmental problem solving?

3. What is meant by *repeatability* in science?

4. What are the five basic steps of the scientific method?

5. Explain why cancer experiments performed on rats may not be accurate predictors of suspected carcinogens' effects on humans.

6. Explain how high levels of nutrients in a body of water lead to death of fish and invertebrates.

7. In the 1950s, W. T. Edmondson formed the hypothesis that treated sewage was adding so much nutrient pollution to Lake Washington that cyanobacteria were multiplying rapidly. This hypothesis made a clear prediction. What was it?

8. Why was it believed that pumping sewage affluent from the Lake Washington area to Puget Sound would not cause the same damage to Puget Sound that Lake Washington had already suffered?

9. What are *global commons*, and what are some examples?

10. Environmental problems tend to be linked to other persistent problems. What are some of these other problems?

Critical-Thinking Questions:

1. Comment on the statement, "There is no absolute truth in science, only varying degrees of uncertainty." Do you agree? If so, how is science able to produce valid conclusions? Or is it?

2. Comment on the statement, "People should not expect no-risk foods, no-risk water, or no-risk anything else." Do you agree? Justify your response.

3. Why do you think politicians so often take a "wait-and-see" approach to environmental problems? Is this a desirable approach?

Data Interpretation:

1. If malaria infects 400 million people per year and kills 2.7 million, what percent of people infected with malaria die from it?

2. If the LD_{50} for cocaine is 17.5 mg/kg, how much cocaine will kill 50% of 140-pound humans? (First, convert mg/kg to mg/lb *or* convert pounds to kilograms.)

Chapter 3

Addressing Environmental Problems, Part II

Study Outline:

I.   Conservation

II.   History of the U.S. Environmental Movement
        A. Protecting Forests
        B. Establishing and Protecting National Parks and Monuments
        C. Conservation in the Mid-20th Century
        D. The Environmental Movement in the Late-20th Century

III.   An Economist's View of Pollution
        A. How Much Pollution Is Acceptable?
        B. Economic Strategies for Pollution Control

IV.   Government and Environmental Policy
        A. Legislative Approaches to Environmental Problems Have Been Effective
            Overall

V.   Case in Point: Old-Growth Forests of the Pacific Northwest
        A. The Environmental Significance of Old-Growth Forests
        B. The Commercial Significance of Old-Growth Forests
        C. Complexities of the Controversy That Were Seldom Portrayed by the Media
        D. The Political Solution to the Controversy
        E. An Economic Case for Preserving Old-Growth Forests

VI.   Ethics, Values, and Worldviews
        A. Worldviews

VII.   Envirobrief: Religious about the Environment

VIII.   Envirobrief: How Green Is Your Campus?

IX.   Envirobrief: Trading Turtle Safety

Key Terms: The following terms are listed in order of appearance in your textbook.

unfunded mandates
conservation
frontier attitude
John James Audubon
Henry David Thoreau
George Perkins Marsh
Theodore Roosevelt
Gifford Pinchot
John Muir
Franklin Roosevelt
Aldo Leopold
Rachel Carson
environmentalists
Gaylord Nelson
Environmental Protection Agency (EPA)
National Environmental Policy Act (NEPA)
environmental impact statements
economics
external cost
marginal cost
marginal cost of pollution
marginal cost of pollution abatement
optimum amount of pollution
emission charge
waste-discharge permits
emission reduction credits (ERCs)
command and control
salvage logging
ethics
values
environmental values
worldview
environmental worldviews
Arne Naess

Multiple Choice:

1. In the four decades following the Civil War, an area the size of Europe was deforested in
    a. the Southeast
    b. Alaska
    c. the Pacific Northwest
    d. the Midwest
    e. the Northeast

2. _____'s book, *Man and Nature*, provided an early (1864) discussion of humans as agents of environmental change.
    a. Henry David Thoreau
    b. John James Audubon
    c. Aldo Leopold
    d. John Muir
    e. George Perkins Marsh

3. When a bill rescinding the president's power to establish forest reserves was passed by Congress in 1907, _____ before signing the bill.
    a. Teddy Roosevelt established many new national forests
    b. Teddy Roosevelt banned logging in national forests
    c. Gaylord Nelson banned logging in national forests
    d. Woodrow Wilson established many new national forests
    e. Woodrow Wilson banned logging in national forests

4. The world's first national park was
    a. Yellowstone
    b. Yosemite
    c. Grand Canyon
    d. Shenandoah
    e. Rocky Mountain

5. The building of the Hetch Hetchy Dam led to the formation of the
    a. Sierra Club
    b. National Park Service
    c. Forest Service
    d. Civilian Conservation Corps
    e. Fish and Wildlife Service

6. The first Earth Day, the signing into law of the National Environmental Policy Act, and the establishment of the Environmental Protection Agency all occurred in
   a. 1950
   b. 1960
   c. 1970
   d. 1980
   e. 1990

7. A loophole in the Clean Air Act of 1977 led to the proliferation of
   a. more-polluting cars
   b. tall smokestacks
   c. lawsuits
   d. industry layoffs
   e. asthma cases

8. An external cost of a product is borne by people who _____ directly involved in the product's market exchange; an external cost usually _____ reflected in the product's price.
   a. are, is
   b. are, is not
   c. are not, is
   d. are not, is not

9. One of the root causes of pollution is the failure to consider _____ in the pricing of goods.
   a. external costs
   b. internal costs
   c. emission charges
   d. marginal costs
   e. unfunded mandates

10. A(n) _____ cost is the additional cost associated with one more unit of something.
    a. marginal
    b. external
    c. internal
    d. optimum
    e. added

11. This graph (Figure 3-8 in *Environment*, 3/e) shows that as amount of pollution increases, the cost of damage
    a. increases
    b. decreases
    c. fluctuates wildly
    d. remains unchanged
    e. drops to zero

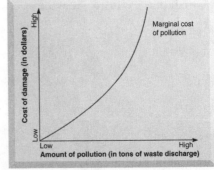

12. This graph (Figure 3-9 in *Environment*, 3/e) shows that as the level of pollution falls, the cost of control
    a. falls
    b. rises
    c. fluctuates wildly
    d. remains unchanged
    e. drops to zero

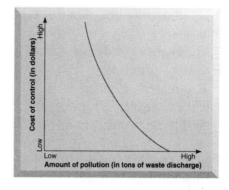

13. To determine the "optimum amount of pollution," an economist uses a(n) _____ diagram that shows the intersection of two _____ curves.
    a. cost-benefit, marginal-cost
    b. marginal-cost, cost-benefit
    c. cost-benefit, external-cost
    d. external-cost, cost-benefit
    e. marginal-cost, waste-discharge

14. Imposing an emission charge on polluters _____ the cost of polluting; generally, the public _____ supportive of such "taxes."
    a. raises, is
    b. raises, is not
    c. lowers, is
    d. lowers, is not
    e. has little effect on, is not

15. The Clean Air Act of 1990 includes a plan to reduce _____ emissions from coal-burning power plants by using tradable permits (emission reduction credits).
    a. carbon dioxide
    b. lead
    c. carbon monoxide
    d. soot
    e. sulfur dioxide

16. Most pollution control efforts in the U.S. have been
    a. voluntary
    b. educational
    c. ineffective
    d. legislative
    e. illegal

17. _____ in developing countries typically have _____ air than cities in developed countries.
    a. Rural areas, dirtier
    b. Cities, dirtier
    c. Rural areas, cleaner
    d. Cities, cleaner
    e. Wilderness areas, dirtier

18. Which of the following has recovered enough to be removed from the federal endangered species list?
    a. dusky seaside sparrow
    b. California condor
    c. black-footed ferret
    d. grizzly bear
    e. bald eagle

19. Since 1970, the level of _____ in the air has dropped 98%.
    a. carbon dioxide
    b. sulfur dioxide
    c. lead
    d. ozone
    e. nitrogen

20. Key players in the spread of fungal spores in forests of the Pacific Northwest are _____; fungi, in turn, help _____ obtain nutrients.
    a. deer, spotted owls
    b. spotted owls, trees
    c. voles, trees
    d. voles, deer
    e. earthworms, earthworms

21. Efforts to protect the endangered spotted owl led to a ban on _____ in 3 million acres (1.2 million hectares) of the Pacific Northwest.
    a. development
    b. logging
    c. hunting
    d. pesticide use
    e. camping and hiking

22. Declines in logging jobs in Oregon *prior* to the spotted-owl controversy were due to
    a. global warming
    b. automation of the timber industry
    c. timber companies' concerns about endangered species
    d. development
    e. decreasing demand for wood

23. This graph (Figure 3-14 in *Environment*, 3/e) shows that *few* remaining old growth forests correspond with _____ public motivation to save them.
    a. high
    b. moderate
    c. low
    d. no

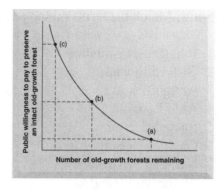

24. Since 1996, Harvard University has hosted interdisciplinary conferences on the _____ of world religions.
    a. economic systems
    b. environmental "sins"
    c. ecological beliefs
    d. environmental fundraising
    e. green architecture

25. In 1998, the World Trade Organization ruled that the U.S. must import shrimp from nations that fail to use technology (called "TEDs") that protects endangered
    a. whales
    b. sharks
    c. sea turtles
    d. tuna
    e. sea urchins

26. Many Central and Eastern Europeans suffer from _____ diseases directly related to pollution.
    a. digestive
    b. liver and kidney
    c. water-borne
    d. vascular
    e. respiratory

Matching: Match the terms on the left with the responses on the right.

_____1. John James Audubon
_____2. Gifford Pinchot
_____3. Antiquities Act
_____4. Aldo Leopold
_____5. Rachel Carson
_____6. Gaylord Nelson
_____7. National Environmental
        Policy Act (NEPA)
_____8. environmental impact
        statement
_____9. economics
_____10. marginal cost of pollution
_____11. marginal cost of pollution
        abatement
_____12. economically optimum
        amount of pollution
_____13. market-oriented strategy
_____14. emission charge
_____15. command and control
_____16. Arne Naess

a. the traditional approach economists take to problems of pollution control
b. monitored by the Council on Environmental Quality
c. promoted wilderness conservation in *A Sand County Almanac* (1949)
d. the added cost, for all present and future society members, of an additional unit of pollution
e. requires the preparation of environmental impact statements
f. the added cost, for all present and future society members, of reducing a given pollutant by one unit
g. serves as a tax on pollution
h. the study of how people use their resources to satisfy their wants
i. represented by the intersection point on a cost-benefit diagram
j. an approach to pollution control that relies on regulatory laws and pollution-level limits
k. alerted the public to pesticide dangers in *Silent Spring* (1962)
l. Senate member who organized the first Earth Day (1970)
m. authorizes the president to set aside national monument sites
n. first head of the U.S. Forest Service; advocated sustainable forestry
o. Norwegian who originated the deep ecology worldview
p. increased interest in American wildlife through his artwork

Fill-In:

1. The passage in 1995 of the _____ (two words) bill takes financial pressure off state and local governments.

2. _____ is the careful management of natural resources.

3. _____ wrote about living on the shore of New England's Walden Pond in the 19th century.

4. John _____ was largely responsible for the establishment of Yosemite and Sequoia National Parks.

5. During the Great Depression, the _____ (three words) was established to employ Americans in jobs that benefited natural resources.

6. In a free market, the price of a good is determined by _____ and _____ .

7. When the cost of environmental harm is not added to the price of products, a market force is generated that _____ (increases or decreases) pollution.

8. The cost of enduring more pollution is measured in terms of _____ (two words).

9. The cost of eliminating pollution is generally measured in terms of giving up _____ .

10. This graph (Figure 3-10 in *Environment*, 3/e) shows that if the amount of pollution is to the *right* of the point of intersection, a polluter will minimize expense by polluting minimize expense by polluting _____ (more or less).

11. Generally, possible environmental disruption or destruction _____ (is or is not) included when economists add up pollution costs.

12. Green taxes are normally set too _____ (high or low), limiting their effectiveness.

13. _____ (three words) are pollution permits that can be bought and sold.

14. The use of ERCs prevents pollution abatement from impeding _____ (two words).

15. The Environmental Protection Agency (EPA) used tradable quotas to reduce _____ levels in gasoline.

16. Since 1970, emissions of sulfur dioxide, carbon monoxide, soot, and motor-vehicle hydrocarbons have _____ (increased or decreased) in the U.S.

17. Most _____ forests in the U.S. are found in the Pacific Northwest and Alaska.

18. In 1994, 3 years after the court-ordered suspension of logging in spotted-owl habitat, Oregon had the _____ (highest or lowest) unemployment rate in 25 years.

19. The U.S. Forest Service spends millions of dollars _____ (more or less) to harvest trees in national forests than it earns from timber sales.

20. Human values are addressed by the branch of philosophy called _____.

21. The _____ (three words) is an environmental worldview that promotes harmony with nature and the belief that humans and all other species have equal worth.

22. The _____ (three words) is a measure of the economy's net production, after a deduction for used-up capital.

23. In an ideal accounting system, the economic cost of _____ (two words) would be subtracted from a firm's contribution to gross domestic product (GDP).

Short Answer:

1. What are some of the uses of national forests?

2. What is an EIS, and what must it include?

3. How did the NEPA (National Environmental Policy Act) revolutionize environmental protection in the U.S.?

4. What are the two major flaws in economists' concept of optimum pollution?

5. Give some examples of "green taxes" in effect or under consideration in some states and countries. Why has the public not been supportive of such taxes?

6. Name several industrial countries that have cut sulfur-dioxide emissions from coal-burning power plants.

7. What is *salvage logging*, and why have forestry scientists and environmentalists opposed it?

8. What is inaccurate about the way the GDP (gross domestic product) is measured?

9. What are some of the signs of environmental devastation in Central and Eastern Europe that came to light after the fall of the Soviet Union?

10. Why did industries and individuals lack motivation to conserve energy in the former Soviet Union?

Critical-Thinking Questions:

1. Economists typically measure pollution costs in terms of property damage, human health costs, and the monetary value of plants and animals killed. What "costs" are missing from such calculations and why? Do you think it is appropriate to assign monetary value to all aspects of the environment? How else might pollution costs be measured, if not in terms of dollars?

2. What do you think is the fairest or most ethical way to resolve conflicts such as the logger/spotted-owl dilemma? Critique the Northwest Forest Plan.

3. Contrast the Western worldview and the deep ecology worldview. Comment on the origin of the Western worldview and the implications of continued adherence to it.

Data Interpretation:

1. By 1897, Michigan sawmills had processed 160 billion board feet of white pine, leaving less than 6 billion board feet standing in the state. What percentage of the total remained in living trees?

2. According to the EPA, the late-1990's cost of complying with federal environmental regulations was $170 billion per year (slightly more than 2% of the U.S. gross domestic product). Given a U.S population of approximately 268 million at that time, what was the cost (borne by industries as well as consumers and taxpayers) per person?

3. Since 1970, annual hydrocarbon emissions from motor vehicles in the U.S. have declined from 10.3 million tons to 5.5 million tons. By what percentage have emissions dropped?

PART TWO:

THE WORLD WE LIVE IN

Chapter 4

Ecosystems and Energy

Study Outline:

Key Terms: The following terms are listed in order of appearance in your textbook.

estuaries
biotic
abiotic
ecology
populations
species
community
ecosystem
landscape ecology
biosphere
atmosphere
hydrosphere
lithosphere
energy
kilojoules (kJ)
kilocalories (kcal)
potential energy
kinetic energy
thermodynamics
closed system
open system
first law of thermodynamics
second law of thermodynamics
entropy
photosynthesis
chlorophyll
cell respiration
energy flow
producers
autotrophs
consumers
heterotrophs
primary consumers
herbivores
secondary consumers
tertiary consumers
carnivores
omnivores
detritus feeders (or detritivores)
detritus
decomposers (or saprotrophs)
food chains
food web
trophic level

pyramid of numbers
pyramid of biomass
biomass
pyramid of energy
gross primary productivity (GPP)
net primary productivity (NPP)
krill

Multiple Choice:

1. The Chesapeake Bay is one of the world's richest
   a. tributaries
   b. estuaries
   c. freshwater systems
   d. swamps
   e. nontidal wetlands

2. The word *ecology* comes from the Greek word for "_____" (*eco*) and the Greek word for "study" (*logy*).
   a. organism
   b. nature
   c. plant
   d. house
   e. life

3. An ecologist who studies rising and falling numbers of striped bass (or "rockfish") in the Chesapeake Bay is a(n) _____ ecologist.
   a. population
   b. community
   c. ecosystem
   d. biosphere
   e. hydrosphere

4. An ecologist who studies predators and prey of the Chesapeake Bay's striped bass, as well as water temperature, oxygen content, salinity, turbidity, and nutrient levels, is a(n) _____ ecologist.
   a. population
   b. community
   c. ecosystem
   d. biosphere
   e. hydrosphere

5. An ecologist who studies interactions among striped bass and their predators and prey in the Chesapeake Bay is a(n) _____ ecologist.
   a. population
   b. community
   c. ecosystem
   d. biosphere
   e. hydrosphere

6. All of Earth's communities of organisms combined make up its
   a. atmosphere
   b. lithosphere
   c. hydrosphere
   d. ecosystem
   e. biosphere

7. _____ is the study of energy and its transformations.
   a. Thermodynamics
   b. Energetics
   c. Kinetics
   d. Ecology
   e. Earth science

8. Through photosynthesis, plants absorb radiant energy and convert it to _____ energy.
   a. heat
   b. mechanical
   c. electrical
   d. nuclear
   e. chemical

9. In all energy transformations, some of the energy is converted to _____ energy.
   a. heat
   b. mechanical
   c. electrical
   d. nuclear
   e. chemical

10. In order to perform photosynthesis, autotrophs require light, water, and
    a. oxygen
    b. carbon dioxide
    c. glucose
    d. chemical energy
    e. atmospheric nitrogen

11. Deer and rabbits are examples of
    a. producers
    b. omnivores
    c. primary consumers
    d. secondary consumers
    e. tertiary consumers

12. Bears, pigs, and meadow voles are examples of
    a. producers
    b. omnivores
    c. detritivores
    d. saprotrophs
    e. autotrophs

13. Snakes, lions, and spiders are examples of
    a. producers
    b. primary consumers
    c. secondary or higher consumers
    d. decomposers
    e. detritivores

14. Earthworms, termites, and crabs are examples of
    a. producers
    b. saprotrophs
    c. decomposers
    d. detritivores
    e. both b. and c.

15. Bacteria and fungi are examples of
    a. producers
    b. saprotrophs
    c. decomposers
    d. detritivores
    e. both b. and c.

16. In a typical food web, herbivores are _____ carnivores.
    a. more abundant than
    b. less abundant than
    c. equally abundant as
    d. at a higher trophic level than
    e. at the same trophic level as

17. On average, there is about a _____% reduction of biomass for each higher trophic level.
    a. 10
    b. 30
    c. 50
    d. 70
    e. 90

18. The open ocean is a(n) _____ ecosystem because of its _____ availability of mineral nutrients.
    a. highly productive, high
    b. somewhat productive, moderate
    c. rather unproductive, limited
    d. extremely unproductive, very limited
    e. none of the above

19. Humans are estimated to use _____% of the annual NPP (net primary productivity) of terrestrial ecosystems.
    a. 10
    b. 25
    c. 40
    d. 55
    e. 70

20. Which of the following ecosystems has the highest NPP?
    a. tropical rain forest
    b. savanna
    c. swamp
    d. boreal forest
    e. agricultural land

21. Which of the following does *not* describe a hydrothermal vent?
    a. tremendous depth
    b. algae at base of food web
    c. tremendous heat
    d. absence of light
    e. hydrogen-sulfide–rich water

22. Baleen whales and squid eat krill, which are
    a. small fish
    b. mollusks
    c. tiny, shrimplike animals
    d. a type of alga
    e. tube worms

23. The _____ whale has not experienced rising numbers since the 1986 global ban on the hunting of large whales.
    a. killer
    b. southern blue
    c. humpback
    d. gray
    e. sperm

24. The greatest threat to the entire Antarctic food web is
    a. reduced numbers of krill
    b. overabundance of algae
    c. declining penguin populations
    d. whale hunting
    e. ozone-layer depletion

25. The greatest biodiversity is found in habitats
    a. that are most productive
    b. that are least productive
    c. that are moderately productive
    d. at high latitudes
    e. with few predators

26. High biodiversity appears to depend on factors that
    a. increase competition
    b. reduce competition
    c. increase predation
    d. reduce predation
    e. maximize productivity

Matching: Match the terms on the left with the responses on the right. Some responses are used more than once.

_g_ 1. population
_e_ 2. species
_n_ 3. community
_d_ 4. ecosystem
_l_ 5. atmosphere
_a_ 6. hydrosphere
_o_ 7. lithosphere
_p_ 8. photosynthesis
_f_ 9. chlorophyll
_m_10. cell respiration
_b_11. producers
_b_12. autotrophs
_k_13. consumers
_k_14. heterotrophs
_j_15. primary consumers
_j_16. herbivores
_q_17. secondary consumers
_q_18. tertiary consumers
_c_19. omnivores
_h_20. detritivores
_i_21. decomposers
_i_22. saprotrophs

a. Earth's supply of water
b. most perform photosynthesis
c. organisms that eat plants and animals
d. a community and its physical environment
e. a group of interbreeding organisms that do not interbreed with members of other groups
f. a pigment that absorbs light (radiant) energy
g. members of a species that live together
h. consumers that eat organic matter
i. microbial heterotrophs that break down dead organic matter
j. organisms that eat producers
k. a term that applies to all animals
l. the gaseous envelope surrounding Earth
m. releases energy stored in organic molecules
n. made up of interacting populations of different species
o. the soil and rock of Earth's crust
p. transforms light energy into chemical energy
q. carnivores

Fill-In:

1. The environment consists of two parts: the _____ (living) environment and the _____ (nonliving, or physical) environment.

2. _____ is the study of the interactions of organisms with one another and with their abiotic environment.

3. One of the characteristics of life is its high degree of _____.

4. _____ (two words) is a sub-discipline of ecology that studies connections among adjacent ecosystems.

5. _____ and _____ are two disciplines that are very important to ecology.

6. Energy expressed in units of work is expressed in _____.

7. Energy expressed in units of heat energy is expressed in _____.

8. Energy can exist as _____ energy (stored energy) or as _____ energy (energy of motion).

9. A(n) _____ system does not exchange energy or matter with its surroundings; a(n) _____ system does.

10. _____ is a measure of disorder or randomness.

11. In cell respiration, $C_6H_{12}O_6 +$ _____ $+ 6H_2O \rightarrow$ _____ $+ 12H_2O +$ energy.

12. The one-way passage of energy through an ecosystem is called _____ (two words).

13. Organic matter such as animal carcasses, leaf litter, and feces is called _____.

14. Producers provide _____ and _____ for the rest of the community.

15. A sequence of energy flow from one organism to the next is called a _____ (two words).

16. A _____ (two words) is a complex of interconnected food chains.

17. Each "link" in a food chain or food web is called a _____ (two words).

18. A pyramid of _____ illustrates the total amount of living material at each successive trophic level.

19. A pyramid of _____ is often expressed as kilocalories per square meter for each trophic level.

20. The _____ (three words) of an ecosystem is the rate at which energy is captured during photosynthesis.

21. The _____ (three words) of an ecosystem represents the rate at which organic matter is incorporated into plant tissues.

Short Answer:

1. What are some of the insects, birds, and invertebrates found in a Chesapeake Bay salt marsh? Why are amphibians uncommon in salt marshes?

2. What forms can energy take?

3. State the first law of thermodynamics.

4. State the second law of thermodynamics.

5. Give the overall equation for photosynthesis.

6. Give examples of consumers that may be *more* numerous than the organisms they feed on. Why are pyramids of numbers of limited usefulness?

7. How is net primary productivity (NPP) calculated?

8. What are the most productive aquatic ecosystems, and why is the open ocean *un*productive?

9. What organisms form the base of food chains in hydrothermal vents, and what are some of the animals they support? Why are there no photosynthetic organisms present?

10. How have seals, penguins, and small baleen whales of Antarctic waters benefited from the decline of large baleen whales?

Critical-Thinking Questions:

1. Relate the second law of thermodynamics to the prediction (by some) that the universe will someday cease to operate. How does the second law of thermodynamics relate also to ecosystem structure?

2. Explain how human use of fossil fuels and fertilizers could be "bad news for global biodiversity."

Data Interpretation:

1. Based on the "10% rule," a person who eats 250 kcal of chicken meat is indirectly consuming how many kcal of corn (the grain fed to the chicken)?

2. Table 4-1 in *Environment*, 3/e, lists the average NPP of agricultural land as 650 grams dry matter/m²/year. What is the average NPP of a temperate deciduous forest and what percent of *this* NPP is represented by the NPP of agricultural land?

3. Suppose the NPP for a boreal forest is measured at 17,034 kcal/m²/year. If respiration by boreal-forest vegetation is 22,155 kcal/m²/year, what is the GPP of this ecosystem?

Chapter 5

Ecosystems and Living Organisms

Study Outline:

I.   Biological Communities
        A. Keystone Species

II.   Interactions Among Organisms
        A. Predation
        B. Symbiosis
        C. Competition

III.   The Ecological Niche
        A. Limiting Factors
        B. Competitive Exclusion and Resource Partitioning

IV.   Species Diversity
        A. Species Diversity, Ecosystem Services, and Community Stability

V.   Evolution: How Populations Change Over Time
        A. Natural Selection

VI.   Succession: How Communities Change Over Time
        A. Primary Succession
        B. Secondary Succession

VII.   Envirobrief: Credible Science in the Biosphere

VIII.   Envirobrief: From the Little Acorn… Comes Lyme Disease?

IX.   Envirobrief: Otters in Trouble

X.   Focus On: The Kingdoms of Life

XI.   Focus On: Coevolution

Key Terms: The following terms are listed in order of appearance in your textbook.

community
resource
ecosystem
keystone species
predation
coevolution
warning coloration
symbiosis
symbionts
mutualism
zooxanthellae
mycorrhizae
commensalism
epiphytes
parasitism
pathogen
competition
intraspecific competition
interspecific competition
ecological niche
habitat
fundamental niche
realized niche
limiting factor
competitive exclusion
resource partitioning
species diversity
ecotone
edge effect
ecosystem services
community stability
evolution
adaptation
natural selection
succession
primary succession

pioneer community
secondary succession

Multiple Choice:

1. The introduction of the Nile perch to East Africa's Lake Victoria contributed to a decline in algae-eating cichlids. The resulting explosion of algae populations has caused
   a. an increase in dissolved oxygen
   b. a decrease in dissolved oxygen
   c. cichlid numbers to return to normal
   d. the Nile perch to decline in number
   e. fishermen's nets to become clogged and useless

2. Which of the following is a result of Lake Victoria's increased turbidity?
   a. declining Nile perch populations
   b. a decrease in dissolved oxygen
   c. increasing cichlid populations
   d. hybridization among cichlid species
   e. an exploding shrimp population

3. An association of different populations of organisms, which interact in a given environment, is a(n)
   a. species
   b. ecosystem
   c. community
   d. ecotone
   e. ecological niche

4. A biological community and its abiotic environment make up a(n)
   a. species
   b. ecosystem
   c. population
   d. ecotone
   e. ecological niche

5. An example of a keystone species, one that greatly affects the nature and structure of its entire ecosystem, is the
   a. red squirrel
   b. gray wolf
   c. jack rabbit
   d. porcupine
   e. turkey vulture

6. _____ consume _____.
   a. Predators, prey
   b. Prey, predators
   c. Symbionts, prey
   d. Parasites, symbionts
   e. Hosts, parasites

7. Milkweeds are toxic to nearly all animals except for ____, which advertise their toxicity with warning coloration.
   a. skunks
   b. goldenrod spiders
   c. katydids
   d. meadow voles
   e. monarch caterpillars and butterflies

8. _____ is a form of symbiosis in which both species benefit.
   a. Mutualism
   b. Parasitism
   c. Intraspecific competition
   d. Predation
   e. Commensalism

9. _____ are mutualistic associations between fungi and plant roots.
   a. Zooxanthellae
   b. Symbionts
   c. Mycorrhizae
   d. Lichens
   e. Alkaloids

10. Epiphytes, common plants in tropical forests, serve as an example of
    a. mutalism
    b. parasitism
    c. intraspecific competition
    d. predation
    e. commensalism

11. A major reason for declining honeybee populations in the U.S. is
    a. parasitism
    b. commensalism
    c. intraspecific competition
    d. interspecific competition
    e. predation

12. A male ruby-throated hummingbird defending its territory against another male ruby-throated hummingbird illustrates
    a. commensalism
    b. predation
    c. mutualism
    d. intraspecific competition
    e. interspecific competition

13. An organism's niche encompasses
    a. its habitat
    b. what it eats and is eaten by
    c. what it competes with
    d. how it interacts with abiotic aspects of its environment
    e. all of the above

14. The brown anole (a lizard), introduced to Southern Florida from Cuba, restricted the range of green anoles; in other words, brown anoles affected the _____ of green anoles.
    a. fundamental niche
    b. realized niche
    c. coevolution
    d. intraspecific competition
    e. dietary needs

15. Competitive exclusion is a result of
    a. coevolution
    b. resource partitioning
    c. predation
    d. interspecific competition
    e. intraspecific competition

16. Resource partitioning can reduce or limit
    a. interspecific competition
    b. fundamental niches
    c. intraspecific competion
    d. a and b
    e. b and c

17. Robert MacArthur's study of five warbler species revealed that _____ helps them avoid _____ competition.
    a. coevolution, interspecific
    b. coevolution, intraspecific
    c. mutualism, interspecific
    d. resource partitioning, interspecific
    e. resource partitioning, intraspecific

18. Generally speaking, species diversity _____ and environmental stress _____ as one moves away from the equator.
    a. increases, increases
    b. increases, decreases
    c. decreases, increases
    d. decreases, decreases
    e. remains unchanged, increases

19. Evolutionary modification that improves the likelihood of survival and reproductive success is called
    a. coevolution
    b. natural selection
    c. adaptation
    d. overproduction
    e. variation

20. Mature, "climax" communities are in a state of continual
    a. equilibrium
    b. flux
    c. decline
    d. overproduction
    e. primary succession

21. In primary succession, the initial pioneer community is likely to have _____ as its most important member.
    a. lichen
    b. moss
    c. crabgrass
    d. ferns
    e. shrubs

22. Primary succession on the island of Krakatoa, devastated by a volcanic eruption in 1883, has
    a. been extremely rapid
    b. resulted in unexpectedly high species diversity
    c. not yet begun
    d. been accelerated by human intervention
    e. been extremely slow

23. Sand-dune succession normally begins with
    a. mosses
    b. ferns
    c. pines
    d. oaks
    e. grasses

24. _____ succession in Yellowstone National Park, following vast wildfires in 1988, has been _____.
    a. Primary, slow
    b. Primary, fast
    c. Secondary, slow
    d. Secondary, fast
    e. Secondary, prevented by severe soil erosion

25. On an abandoned farm on the North Carolina Piedmont, hardwoods typically replace pines because hardwood seedlings are more tolerant of
    a. acidic soils
    b. direct sunlight
    c. drought
    d. mycorrhizae
    e. shade

26. Bacteria have a(n) _____ cell structure; they _____ a nuclear envelope and other internal cell membranes.
    a. prokaryotic, have
    b. prokaryotic, lack
    c. eukaryotic, have
    d. eukaryotic, lack
    e. symbiotic, have

27. Members of the kingdom _____ frequently live in anaerobic and other harsh environments.
    a. Archaebacteria
    b. Eubacteria
    c. Protista
    d. Fungi
    e. Plantae

28. Eukaryotic organisms include
    a. protists
    b. fungi
    c. plants
    d. animals
    e. all of the above

29. Algae, protozoa, and slime molds are examples of
    a. bacteria
    b. fungi
    c. protists
    d. plants
    e. none of the above

30. Flowering plants and their animal pollinators serve as an example of
    a. coevolution
    b. commensalism
    c. interspecific competition
    d. parasitism
    e. resource partitioning

Matching: Match the terms on the left with the responses on the right.

_e_ 1. killer whales   a. secrete enzymes and absorb predigested food
_b_ 2. goldenrod spiders   b. predators that use camouflage
_k_ 3. honeybees   c. temperature is a limiting factor that restricts their range
_m_ 4. skunks   d. dominant in the second year of secondary succession (in
_f_ 5. zooxanthellae   North Carolina, for example)
_g_ 6. mites   e. predators that hunt in packs
_l_ 7. crown gall bacteria   f. algae that live symbiotically within corals
_c_ 8. ring-neck pheasants   g. parasites of honeybees
_h_ 9. mosses   h. play a role in primary succession
_d_ 10. horseweed   i. eaten by pollinators
_j_ 11. broomsedge   j. drought-tolerant plant common in the third year of
_a_ 12. fungi   secondary succession (in North Carolina, for example)
_i_ 13. nectar and pollen   k. gain protection by living in groups; also use warning
        coloration
     l. plant pathogens
     m. use warning coloration along with an acrid chemical
        spray

Fill-In:

1. A species that plays a vital role in determining the nature and structure of its entire
   ecosystem is called a _____ (two words).

2. The interdependent evolution of two interacting species is called _____.

3. _____, found in tobacco, is a common ingredient in many
   insecticides.

4. _____ is an intimate relationship between members of different
   species.

5. The partners of a symbiotic relationship are called _____.

6. _____ are algae that live symbiotically inside coral cells.

7. _____ is a type of symbiosis in which one member benefits and the other is neither harmed nor helped.

8. _____ is a type of symbiosis in which one member benefits and the other is harmed.

9. A disease-causing (and sometimes death-causing) parasite is called a _____.

10. _____ occurs when individuals try to use the same essential resource.

11. Competition between members of different species is called _____ competition.

12. The local environment in which an organism lives is its _____.

13. An organism's *potential* ecological niche is its _____ (two words).

14. An environmental resource that restricts an organism's niche is called a _____ (two words).

15. The number of species in a given community is referred to as _____ (two words).

16. An _____ is a transitional zone where different communities meet.

17. The _____ (two words) is the change in species composition found at ecotones.

18. Species diversity promotes _____ (two words)—the ability of a community to withstand environmental disturbances—and makes ecosystems better able to supply _____ (two words), such as clean water and fertile soil.

19. _____ is the genetic change in populations that occurs over time.

20. Darwin's mechanism of evolutionary change is called _____ (two words).

21. The process of community development, which involves a series of species replacements, is called _____.

22. The initial community that develops during primary succession is called the _____ (two words).

23. _____ succession occurs in an environment that has not previously supported life.

24. _____ succession occurs after a disturbance destroys existing vegetation; soil, however remains.

Short Answer:

1. Why was the Nile perch introduced to Lake Victoria, and what effects has this introduction had on native species?

2. What are the three main roles organisms play in community life? Give an example of an organism that plays each role.

3. Explain how fig trees act as keystone species in tropical rain forests.

4. Explain how animals that live in groups, such as antelope, gain protection from predators.

5. What happened when Russian biologist G. F. Gause tried to grow two species of *Paramecium* in the same test tube? Why? What ecological phenomenon does this experiment demonstrate?

6. Why is it that species diversity is self-perpetuating?

7. How is species diversity related to geographical isolation? Why?

8. What is the effect on species diversity when one species has clear dominance within a community?

9. What four observations about the natural world form the basis of Darwin's evolutionary mechanism called *natural selection*?

10. What is the typical sequence of secondary succession on abandoned farmland in North Carolina?

11. Explain the connections among white-footed mice, oaks, gypsy moths, deer, and the spread of Lyme disease.

12. Explain how declining fish populations are related to kelp forest decline off the Alaskan coast.

13. List the six kingdoms of life in the six-kingdom classification system.

Critical-Thinking Questions:

1. Explain why the removal of a species' competitors or predators from a community might *not* be beneficial to the species. Give examples.

2. The text mentions the coexistence of native and introduced fish in Florida as well as plants competing in the same location as examples of situations that seem to contradict the competitive exclusion principle. Assuming the competitive exclusion principle is valid, how might these examples involve differences in the realized niches of competing species?

Data Interpretation:

1. How does the combined area of Canada and the continental U.S. compare, percent-wise, to Columbia, Ecuador, and Peru, in terms of native plant species?

2. Darwin used elephants as his example of reproductive potential. Suppose each pair of breeding elephants produces 4, not 6, offspring. Starting with 1 pair, how many elephants would there be in 10 generations?

Chapter 6

Ecosystems and the Physical Environment

Study Outline:

I.   The Cycling of Materials within Ecosystems
        A. The Carbon Cycle
        B. The Nitrogen Cycle
        C. The Phosphorus Cycle
        D. The Hydrologic Cycle

II.   Solar Radiation
        A. Temperature Changes with Latitude
        B. Temperature Changes with Season

III.   The Atmosphere
        A. Layers of the Atmosphere
        B. Atmospheric Circulation
        C. Surface Winds

IV.   The Global Ocean
        A. Patterns of Circulation in the Ocean
        B. Vertical Mixing of Ocean Water
        C. The Ocean Interacts with the Atmosphere

V.   Weather and Climate
        A. Precipitation
        B. Rain Shadows
        C. Tornadoes
        D. Tropical Cyclones

VI.   Internal Planetary Processes
        A. Volcanoes
        B. Earthquakes

VII.   Mini-Glossary of the Nitrogen Cycle

VIII.   Envirobrief: Albedo and Climate Change

IX.   Envirobrief: Humans Change Earth's Rotation

X.   Mini-Glossary of Köppen's Climate Zones

XI.   Summary with Selected Key Terms

Key Terms: The following terms are listed in order of appearance in your textbook.

Gaia hypothesis
geophysiology
negative feedback loop
biogeochemical cycles
carbon cycle
fossil fuels
combustion
nitrogen cycle
nitrogen fixation
nitrogenase
nodules
heterocysts
nitrification
assimilation
ammonification
denitrification
nitrogen oxides
photochemical smog
acid deposition
phosphorus cycle
hydrologic cycle
transpiration
estuaries
runoff
watershed
groundwater
albedo
troposphere
stratosphere
mesosphere
thermosphere
exosphere
winds
Coriolis effect

prevailing winds
polar easterlies
westerlies
trade winds
currents
gyres
density
El Niño-Southern Oscillation (ENSO)
upwells
weather
climate
rain shadow
tornado
tropical cyclones
plate tectonics
plate boundary
subduction
magma
lava
hot spot
seismic waves
faults
focus
epicenter

Multiple Choice:

1. The Gaia hypothesis attempts to explain how
   a. organisms are affected by the environment
   b. variations occur in the composition of the atmosphere and seas
   c. solar radiation is stored and used to do biological work
   d. organisms modify the environment to keep it habitable for life
   e. photochemical smog affects local climate

2. Photosynthesis is an important part of the _____ cycle.
   a. carbon
   b. hydrologic
   c. nitrogen
   d. phosphorus
   e. sulfur

3. A great deal of _____ is stored in limestone and in the wood of trees.
    a. nitrogen
    b. water
    c. carbon
    d. phosphorus
    e. potassium

4. Since 1850, combustion of wood and fossils fuels has been increasing the amount of
    _____ in the atmosphere.
    a. water vapor
    b. oxygen
    c. carbon dioxide
    d. nitrogen
    e. methane

5. The atmosphere consists mostly of which gas?
    a. nitrogen
    b. oxygen
    c. water vapor
    d. carbon dioxide
    e. hydrogen

6. Which cycle depends most heavily on the action of bacteria?
    a. hydrologic
    b. nitrogen
    c. carbon
    d. phosphorus
    e. magnesium

7. *Rhizobium* and filamentous cyanobacteria are important to which part of the nitrogen
    cycle?
    a. nitrogen fixation
    b. nitrification
    c. assimilation
    d. ammonification
    e. denitrification

8. Anaerobic conditions are required by bacteria involved in _____ and _____.
    a. nitrogen fixation, nitrification
    b. nitrification, assimilation
    c. assimilation, ammonification
    d. ammonification, denitrification
    e. nitrogen fixation, denitrification

9. Conversion of ammonia and ammonium to nitrite, and then to nitrate, is called
   a. nitrogen fixation
   b. nitrification
   c. assimilation
   d. ammonification
   e. denitrification

10. _____ contribute(s) to photochemical smog, acid deposition, global warming, and ozone depletion.
    a. Nitrates in fertilizers
    b. Phosphates in fertilizers
    c. Releases of carbon dioxide from combustion of fossil fuels
    d. Nitrous oxides from automobile exhaust
    e. Photosynthesis

11. Feedlot wastes and sewage are major sources of _____ pollution in rivers and lakes.
    a. sulfur
    b. carbon dioxide
    c. phosphate
    d. mercury
    e. lead

12. Transpiration, an important part of the _____ cycle, is performed by _____.
    a. carbon, plants
    b. nitrogen, bacteria
    c. carbon, bacteria
    d. hydrologic, plants
    e. phosphorus, animals

13. Which of the following have *low* albedos?
    a. oceans and forests
    b. asphalt surfaces
    c. glaciers and ice sheets
    d. all of the above
    e. a and b only

14. Seasons are determined primarily by
    a. changes in air density
    b. changes in solar energy output
    c. Earth's inclination on its axis of rotation
    d. changes in Earth's albedo
    e. atmospheric circulation

15. The evolution of photosynthesis led to the accumulation of _____ in the atmosphere.
    a. carbon dioxide
    b. oxygen
    c. nitrogen
    d. methane
    e. CFCs

16. The layer of the atmosphere closest to Earth's surface is the
    a. troposphere
    b. stratosphere
    c. mesosphere
    d. thermosphere
    e. exosphere

17. The outermost layer of the atmosphere is the
    a. troposphere
    b. stratosphere
    c. mesosphere
    d. thermosphere
    e. exosphere

18. Two layers of the atmosphere are characterized by rising temperature with increasing altitude. They are the _____ and the _____.
    a. troposphere, stratosphere
    b. stratosphere, mesosphere
    c. mesosphere, thermosphere
    d. thermosphere, exosphere
    e. stratosphere, thermosphere

19. _____ are winds that blow in the mid-latitudes (from the southwest in the Northern Hemisphere, from the northwest in the Southern Hemisphere).
    a. Prevailing winds
    b. Polar easterlies
    c. Westerlies
    d. Trade winds
    e. Currents

20. _____ are tropical winds (from the northeast in the Northern Hemisphere, from the southeast in the Southern Hemisphere).
    a. Prevailing winds
    b. Polar easterlies
    c. Westerlies
    d. Trade winds
    e. Currents

21. This figure (Figure 6-15 in *Environment*, 3/e) shows the sinking of _____ water in the _____, which helps drive the ocean convevor belt.
    a. warm, North Atlantic
    b. cold, North Atlantic
    c. warm, North Pacific
    d. cold, North Pacific
    e. salty, Indian Ocean

22. Ocean water that is _____ is denser than _____ water.
    a. warmer or saltier, colder or less salty
    b. colder or saltier, warmer or less salty
    c. warmer or less salty, colder or saltier
    d. colder or less salty, warmer or saltier
    e. none of the above

23. An El Niño-Southern Oscillation (ENSO) event occurs when trade winds _____ and _____ Pacific waters spread eastward to South America.
    a. strengthen, warm
    b. weaken, warm
    c. strengthen, nutrient-rich
    d. weaken, cool
    e. strengthen, cool

24. An ENSO _____ upwelling along the west coast of South America, which causes dramatic _____ in many fish populations.
    a. suppresses, declines
    b. suppresses, increases
    c. strengthens, declines
    d. strengthens, increases
    e. strengthens, fluctuations

25. _____ has the most profound impact on the distribution and kinds of terrestrial organisms.
    a. Wind
    b. Air pressure
    c. Hurricane frequency
    d. Precipitation
    e. The Coriolis effect

26. As air rises, it _____, and its water-holding capacity _____.
    a. warms, increases
    b. warms, decreases
    c. cools, increases
    d. cools, decreases
    e. cools, remains unchanged

27. Which country has the most tornadoes?
    a. China
    b. Kenya
    c. the United States
    d. Brazil
    e. India

28. The _____ Ocean is getting _____; this change is an effect of plate tectonics.
    a. Indian, smaller
    b. Indian, larger
    c. Pacific, smaller
    d. Pacific, larger
    e. Atlantic, larger

29. The Hawaiian Islands formed at a
    a. subduction zone
    b. spreading center
    c. hot spot
    d. plate boundary
    e. continental shelf

30. The moment magnitude scale is used to measure
    a. earthquakes
    b. tornado winds
    c. hurricane winds
    d. ocean currents
    e. trade winds

31. One of North America's most geologically active places is in _____, where the San Andreas fault occurs.
    a. western Canada
    b. California
    c. the upper Midwest
    d. New England
    e. the Carolinas

32. Melting glaciers increase oceans' surface area, which _____ planetary albedo and accelerates global _____.
    a. raises, cooling
    b. raises, warming
    c. lowers, cooling
    d. lowers, warming
    e. stabilizes, cooling

Matching: Match the terms on the left with the responses on the right. Some responses are used more than once.

_____ 1. homeostatic mechanism
_____ 2. geophysiology
_____ 3. limestone
_____ 4. cell respiration
_____ 5. fossil fuels
_____ 6. combustion of wood as fuel
_____ 7. heterocysts in filamentous cyanobacteria
_____ 8. *Rhizobium* and *Nitrobacter*
_____ 9. nitrogen
_____10. phosphorus
_____11. transpiration
_____12. albedo
_____13. Coriolis effect
_____14. El Niño
_____15. La Niña
_____16. focus
_____17. epicenter

a. determines how much sunlight is reflected from a surface
b. found in proteins and nucleic acids
c. part of the hydrologic cycle
d. influences ocean currents and wind directions
e. part of the carbon cycle
f. involves a decrease in surface temperatures in the East Pacific
g. a site of nitrogen fixation
h. involves an increase in surface temperatures in the East Pacific
i. helps maintain a steady state or a constant environment
j. an earthquake's surface location, directly above its site of origin
k. make nitrogen available to plants
l. found in nucleic acids but not in proteins
m. field of study based on the Gaia hypothesis
n. site of an earthquake's origin

Fill-In:

1. A _____ (two words) loop between organisms and the abiotic environment helps regulate and stabilize Earth's temperature.

2. Cycles of matter that involve biological, geological, and chemical interactions are called _____ cycles.

3. With respect to matter, Earth is considered a _____ system.

4. Only about _____% of the atmosphere is carbon dioxide.

5. When burning, or _____ occurs, organic molecules are _____ (combined with oxygen).

6. Table 6-1 in *Environment*, 3/e, shows that most of Earth's carbon is found in _____.

7. Nitrogen-fixing bacteria use an enzyme called _____ to split atmospheric nitrogen molecules and combine nitrogen atoms with hydrogen.

8. Nitrogen-fixing bacteria live inside root swellings, called _____, found on plants (such as clover, peas, and beans) collectively called _____.

9. Conversion of ammonia or ammonium to nitrate is called _____.

10. The absorption and utilization of nitrate, ammonia, or ammonium by plants is called _____.

11. Conversion of biological nitrogen compounds, such as protein subunits, into ammonia and ammonium ions is called _____.

12. Conversion of nitrate to nitrogen gas, which returns to the atmosphere, is called _____.

13. _____ oxides are a necessary ingredient in the reactions that produce _____ (two words), a mixture of several air pollutants that acts as an eye and respiratory-tract irritant; these oxides also cause _____ (two words), which lowers the pH of waterways and soils.

14. Water running over rocks helps move inorganic _____ molecules into the soil, where they can be taken up by plants.

15. An estuary, such as the Chesapeake Bay, receives runoff from its entire _____ (area of land that drains into it).

16. Water that percolates downward through soil and rock becomes _____.

17. The _____ is the layer of the stratosphere that contains a protective layer of ozone.

18. The _____ is the layer of the atmosphere directly above the stratosphere.

19. Gases in the _____ (a layer of the atmosphere) absorb x-rays and short-wave ultraviolet radiation.

20. Circular ocean currents are called _____.

21. After an ENSO, there is often a _____ (two words) event, which has difficult-to-predict weather effects.

22. Weather changes _____ (slowly or quickly); climate changes more _____ (slowly or quickly) than weather.

23. The most important factors that influence an area's climate are _____ and _____.

24. The dry side of a mountain range, away from the prevailing wind, is called a _____ (two words).

25. Giant, rotating tropical storms are called _____ (two words).

26. Even more destructive than their winds are typically the _____ (two words) of tropical cyclones.

27. The Western Hemisphere's deadliest hurricane in at least 200 years struck Central America's Atlantic coast in 1998; this hurricane carried the name _____.

28. The movement of crustal plates, which may collide or slide or separate along crustal _____, is called _____ (two words).

29. _____ is the process in which one crustal plate descends under another.

30. Underground molten rock is called _____; when it reaches the surface, it is called _____.

31. Earthquake energy is released as _____ (two words); most earthquakes occur along _____, or fractures in the crust.

32. The construction of many large reservoirs in the Northern Hemisphere may have sped up Earth's _____.

Short Answer:

1. According to proponents of the Gaia hypothesis, why has the Earth not gotten hotter in response to the sun's increase in energy output?

2. Why has the level of carbon dioxide in the atmosphere increased significantly in the last half of the 20th century?

3. What are the five steps of the nitrogen cycle, and which involve the action of bacteria?

4. What are the sources and consequences of excess nitrogen in bodies of water?

5. What role do seabirds play in the phosphorus cycle?

6. The fraction of incoming solar radiation that is absorbed (as opposed to reflected) plays what roles on Earth?

7. Why is Earth warmer near the equator than it is near the poles?

8. Starting closest to Earth's surface, list the five layers of the atmosphere.

9. What are prevailing winds, and what are their names?

10. What is the ocean conveyor belt, and what were the consequences when it reorganized 11,000 to 12,000 years ago?

11. What were the effects of the 1997–98 ENSO?

12. How does La Niña typically affect the U.S.?

13. What are the six climate zones? Determine from Figure 6-18 in *Environment*, 3/e, which climate zone you live in.

Critical-Thinking Questions:

1. How does photosynthesis both contribute to and combat global warming?

2. Why does temperature decrease with increasing elevation? (In other words, why is a mountain colder at the top than at the bottom?)

Data Interpretation:

1. If nitrogen-fixing bacteria require the energy of 12 grams of glucose to fix 1 gram of nitrogen, how many grams of glucose do they require to fix 4 ounces of nitrogen? (See Appendix IV.)

2. If 22.4 kilograms of fertilizer are used for every person per year, how many kilograms of fertilizer per year support a country with a population the size of Guatemala's? (See the *World Population Data Sheet* in the back of your text.) Express your answer also in pounds. (See Appendix IV.)

3. What percentage of incoming radiation is absorbed by a given area of ocean? Do oceans have high or low albedos?

4. How much more energy is released by a magnitude 8 earthquake than by a magnitude 5 earthquake?

Chapter 7

Major Ecosystems of the World

Study Outline:

I.   Earth's Major Biomes
        A. Tundra: Cold Boggy Plains of the Far North
        B. Taiga: Evergreen Forests of the North
        C. Temperate Rain Forest: Lush Temperate Forests
        D. Temperate Deciduous Forest: Broad-Leaved Trees that Shed Their Leaves
        E. Grasslands: Temperate Seas of Grass
        A. Chaparral: Thickets of Evergreen Shrubs and Small Trees
        B. Deserts: Arid Life Zones
        C. Savanna: Tropical Grasslands
        D. Tropical Rain Forests: Lush Equatorial Forests

II.   Aquatic Ecosystems

III.   Freshwater Ecosystems
        A. Rivers and Streams: Flowing-Water Ecosystems
        B. Lakes and Ponds: Standing-Water Ecosystems
        C. Marshes and Swamps: Freshwater Wetlands

IV.   Estuaries: Where Freshwater and Saltwater Meet

V.   Marine Ecosystems
        A. The Intertidal Zone: Transition Between Land and Ocean
        B. The Benthic Environment: Seagrass Beds, Kelp Forests, and Coral Reefs
        C. The Neritic Province: Shallow Waters Close to Shore
        D. The Oceanic Province: Most of the Ocean
        E. The Impact of Human Activities on the Ocean

VI.   Case In Point: The Everglades
        A. How the Everglades Was Almost Destroyed
        B. Restoration of the Everglades
        C. Follow-up

Key Terms: The following terms are listed in order of appearance in your textbook.

biome
tundra (or arctic tundra)
alpine tundra
permafrost
taiga (or boreal forest)
temperate rain forest
temperate deciduous forest
leach
temperate grasslands
mediterranean climates
chaparral
deserts
savanna
tropical rain forests
salinity
plankton
phytoplankton
zooplankton
nekton
benthos
flowing-water ecosystem
standing-water ecosystem
littoral zone
limnetic zone
profundal zone
thermal stratification
thermocline
fall turnover
spring turnover

blooms
freshwater wetlands
ecosystem services
estuary
salt marshes
mangrove forests
intertidal zone
benthic environment
abyssal zone
sea grasses
kelps
coral reefs
zooxanthellae
fringing reef
atoll
barrier reef
neritic province
euphotic zone
oceanic province
marine snow

Multiple Choice:

1. The arctic tundra is a biome with _____ precipitation and _____ days during the short growing season.
    a. much, long
    b. much, short
    c. little, long
    d. little, short
    e. moderate, hot

2. The tundra's permafrost _____ the growth of most woody plants and causes _____ soil during the summer.
    a. stimulates, nutrient-poor
    b. prevents, nutrient-rich
    c. stimulates, dry
    d. prevents, waterlogged
    e. prevents, nutrient-poor

3. This biome has low species diversity, no reptiles, and few amphibians, but it is home to many insects and birds in summer:
    a. taiga
    b. temperate rain forest
    c. arctic tundra
    d. savanna
    e. temperate deciduous forest

4. The dominant plants of the taiga are
    a. conifers
    b. aspen and birch
    c. mosses and lichens
    d. grasses and sedges
    e. epiphytes

5. The tundra is characterized by
    a. acidic, mineral-poor soils
    b. high annual precipitation
    c. numerous pond and lakes
    d. all of the above
    e. a. and c. only

6. Cool summers, mild winters, huge evergreens, and pronounced ecosystem complexity are found in the
    a. tropical rain forest
    b. savanna
    c. chaparral
    d. temperate deciduous forest
    e. temperate rain forest

7. Oak, hickory, and beech (broad-leaved hardwoods) typically dominate in _____ forests of the _____ U.S.
    a. temperate rain, northwestern
    b. tropical rain, southeastern
    c. boreal, northern
    d. temperate deciduous, northeastern and mideastern
    e. boreal, northeastern and mideastern

8. Worldwide, _____ were/was among the first biomes to be cleared for agriculture.
    a. the taiga
    b. temperate deciduous forests
    c. the savanna
    d. the chaparral
    e. tropical rain forests

9. Grasslands are characterized by
   a. mild winters
   b. predictable rainfall
   c. periodic wildfires
   d. all of the above
   e. none of the above

10. The breadbaskets of the world are found in former
   a. tropical rain forests
   b. temperate rain forests
   c. temperate grasslands
   d. boreal forests
   e. deserts

11. Chaparral plants are dormant during the _____, and many grow _____ in the months following a fire.
   a. winter, most
   b. winter, least
   c. summer, most
   d. summer, least
   e. spring, most

12. Soils low in organic matter and high in mineral content, particularly salts, are found in
   a. deserts
   b. the chaparral
   c. tropical rain forests
   d. the tundra
   e. temperate rain forests

13. Evergreen broad-leaved trees with buttresses and shallow roots, growing in mineral-poor soil, are found in
   a. tropical rain forests
   b. the taiga
   c. temperate rain forests
   d. temperate deciduous forests
   e. the chaparral

14. Which of the following would you *not* expect to find in a tropical rain forest?
   a. lianas
   b. parrots
   c. sloths
   d. epiphytes
   e. giraffes

15. Fishes, turtles, and whales are assigned to the ecological category (of aquatic ecosystems) called
    a. plankton
    b. phytoplankton
    c. zooplankton
    d. benthos
    e. nekton

16. The littoral, limnetic, and profundal zones are found in
    a. rivers and streams
    b. lakes
    c. marshes and swamps
    d. the open ocean
    e. estuaries

17. Aquatic vegetation, frogs, turtles, insect larvae, many fishes, and high productivity are associated with the _____ zone of a lake or pond.
    a. littoral
    b. limnetic
    c. profundal
    d. thermocline
    e. estuarine

18. Thermal stratification of a temperate lake is
    a. greatest in spring
    b. greatest in summer
    c. greatest in fall
    d. greatest in winter
    e. unvaried throughout the year

19. Which of the following is *not* a reason estuaries are among the world's most fertile ecosystems?
    a. abundant nutrients
    b. light penetration of shallow water
    c. constant water temperature
    d. good water circulation
    e. abundant aquatic plants

20. Mangrove forests play an important role
    a. in controlling flood damage from storms
    b. as nurseries for marine species
    c. as nesting sites for birds
    d. in preventing coastal erosion
    e. all of the above

21. Ocean tides are caused by
    a. the Earth's tilt on its axis
    b. wind and local topography
    c. the gravitational pull of the sun
    d. the gravitational pull of the moon
    e. both c. and d.

22. An important food for ducks, geese, green turtles, sea urchins, and certain fishes are
    a. kelps
    b. corals
    c. algae
    d. sea grasses
    e. sponges

23. A(n) _____ ecosystem is the most diverse marine environment.
    a. coral reef
    b. intertidal zone
    c. abyssal zone
    d. benthic
    e. deep-sea

24. Organic debris that drifts from sunlit to deep-water regions is called
    a. plankton
    b. zooxanthellae
    c. kelp
    d. marine snow
    e. pelagic waste

25. The Everglades' two major problems are _____ and _____.
    a. too little water, nutrient pollution
    b. too much water, nutrient pollution
    c. too little water, the frequency and severity of hurricanes
    d. toxic wastes, the frequency and severity of hurricanes
    e. toxic wastes, sewage wastes

26. As one moves from the top to the bottom of a snow- or ice-capped mountain, one is
    likely to encounter the following plants in this order:
    a. grasses, deciduous trees, coniferous trees
    b. deciduous trees, grasses, coniferous trees
    c. grasses, coniferous trees, deciduous trees
    d. coniferous trees, grasses, deciduous trees
    e. deciduous trees, coniferous trees, grasses

27. Alpine tundra typically has _____ precipitation and _____ intense sunlight than arctic tundra does.
    a. more, more
    b. more, less
    c. less, more
    d. less, less

28. Which of the following is *not* a fire-adapted ecosystem?
    a. African savanna
    b. temperate deciduous forest
    c. California chaparral
    d. North American grasslands
    e. pine forests of the southern U.S.

29. Based on Figure 7-1 in *Environment*, 3/e, which biome covers the most area?
    a. chaparral
    b. tropical rain forests
    c. temperate grasslands
    d. temperate deciduous forests
    e. taiga

Matching I: Biomes
Match the phrases on the left with the responses on the right. Some responses are used more than once.

_____ 1. its animals tend to be small; amphibians are uncommon
_____ 2. wildlife includes lemmings, snowshoe hares; plants include mosses, grasses
_____ 3. home to great herds of hoofed mammals
_____ 4. may form sod and highly productive agricultural soils
_____ 5. dominated by broad-leaved trees that shed their leaves
_____ 6. characterized by permafrost
_____ 7. typically have ancient, mineral-poor soils
_____ 8. includes some old-growth, evergreen forests of the Pacific Northwest
_____ 9. include North America's rarest biome
_____ 10. home of moose, lynx, mink; amphibians and reptiles are rare
_____ 11. suffers overgrazing, especially in Africa
_____ 12. boreal forest
_____ 13. once contained puma, wolves, bison
_____ 14. covers large parts of California; fires are fairly common
_____ 15. biome with the greatest species diversity

a. tundra
b. taiga
c. temperate rain forest
d. temperate deciduous forest
e. grasslands
f. chaparral
g. deserts
h. savanna
i. tropical rain forests

Matching II: Aquatic Ecosystems
Match the phrases on the left with the responses on the right. Some responses are used more than once.

_____ 1. vital shallow-water habitats for migratory waterfowl
_____ 2. a productive yet stressful ecosystem
_____ 3. coastal waters where fresh water and saltwater mix
_____ 4. depend heavily on the land for their energy, much of which comes from detritus
_____ 5. open ocean to a depth of 200 m; where swimmers and floaters live
_____ 6. characterized by zonation: littoral, limnetic, profundal
_____ 7. home of giant squid and other predators; also filter feeders, scavengers
_____ 8. where sea grasses, kelps, and coral reefs are found
_____ 9. where salt marshes are found
_____ 10. standing-water ecosystems characterized by thermal stratification
_____ 11. have gained some legal protection from development
_____ 12. the "tropical equivalent of salt marshes"
_____ 13. the "deep-sea"; some of its animals have light-producing organs
_____ 14. contains the ocean's euphotic zone
_____ 15. the ocean floor

a. rivers and streams
b. lakes and ponds
c. freshwater wetlands
d. estuaries
e. mangrove forests
f. intertidal zone
g. benthic environment
h. neritic province
i. oceanic province

Fill-In:

1. A _____ encompasses many interacting ecosystems.

2. Chaparral vegetation is associated with so-called _____ climates.

3. Joshua trees, yuccas, and giant saguaro are found in _____ (which biome?) of North America.

4. The _____ is a tropical grassland with scattered clumps of low trees.

5. The concentration of dissolved salts in an aquatic ecosystem determines its _____, which affects the kinds of organisms present.

6. In aquatic ecosystems, _____ are free-floating, photosynthetic cyanobacteria and algae; _____ are tiny nonphotosynthetic organisms, such as protozoa.

7. _____ are bottom-dwelling organisms in aquatic ecosystems.

8. _____-water ecosystems are characterized by zonation; _____-water ecosystems are characterized by major changes between source and mouth.

9. The _____ zone lies beneath the _____ zone of a large lake.

10. In this figure (Fig. 7-15a in *Environment*, 3/e), the horizontal portion of the graphed line represents the _____ of a temperate lake.

(a)

11. Falling temperatures in autumn cause a mixing of lake waters; this event is called the _____ (two words); the _____ (two words) occurs when ice melts and the surface water sinks.

12. Population explosions of algal and cyanobacterial populations are called _____.

13. A freshwater ecosystem dominated by grasslike plants is a _____; a freshwater ecosystem dominated by trees or shrubs is a _____.

14. Crucial roles that ecosystems play are called _____ (two words).

15. _____ are subject to significant daily, seasonal, and annual fluctuations in temperature, salinity, and depth of light penetration.

16. An area of shoreline between low and high tides is called an _____ (two words).

17. The deepest part of the benthic environment is called the _____ (two words).

18. Symbiotic algae that live and photosynthesize within coral tissues are called _____.

19. The most common type of coral reef is a _____ (two words), which is attached to the shore of a volcanic island or continent; a _____ (two words) is separated from land by a lagoon of open

water; and an _____ is a circular coral reef that forms on top of a submerged volcano and surrounds a central lagoon.

20. Shallow waters close to shore make up an ocean's _____ province.

21. Waters at depths greater than 200 m make up an ocean's _____ province, the largest marine environment.

22. Southern Florida's vast area of sawgrass wetlands is called the _____.

23. Oceanic warming is blamed for recent episodes of coral _____, in which corals expel their symbiotic _____.

Short Answer:

1. What are some of the physical and behavioral adaptations of the black-tailed prairie dog that suit it to its Great Plains niche?

2. What abiotic factors help define and delineate biomes?

3. Name some animals commonly found (or once found) in temperate grasslands of the U.S.

4. In what ways have humans altered American deserts?

5. Where are the world's savannas found?

6. Name and describe the three distinct layers of vegetation found in a tropical rain forest.

7. What abiotic factors most strongly affect aquatic ecosystems?

8. In what ways do human activities damage coral reefs?

9. In what ways do human activities harm the oceans?

10. Describe the three-part plan to partially restore the Everglades.

11. Give some examples of ecosystem services and ecosystem goods. In a study of 16 biomes, which proved most valuable (from a monetary perspective)?

Critical-Thinking Questions:

1. Why have the world's oceans been particularly prone to pollution and overharvesting? How is the answer to the former question related to the particular difficulty of reducing problems of pollution and overharvesting? How might enforcement of regulations be made fair and effective?

2. Historically, humans have suppressed fires started by lightning, arson, or human carelessness. In what ways can ecosystems be harmed by this practice? Do you think it is better to "let Nature take its course" or to intervene when lightning starts fires that don't threaten human structures?

Data Interpretation:

1. Deserts and rain forests are biomes with opposite extremes of precipitation. In contrasting precipitation amounts, assume a given desert receives 15 cm of rainfall per year and a tropical rain forest receives 350 cm. Convert both amounts to inches. (See Appendix IV.) How many more inches of rainfall per year does the rain forest receive?

2. Before human encroachment and engineering began to alter the Everglades, the ecosystem was 80 km wide and 160 km long. How many square kilometers was the entire area of the Everglades? Express this area as square miles. (See Appendix IV.)

PART THREE:

A CROWDED WORLD

Chapter 8

Understanding Population Growth

Study Outline:

Key Terms: The following terms are listed in order of appearance in your textbook.

population ecology
population density
birth rate (*b*)
death rate (*d*)
growth rate (*r*)
natural increase
immigration (*i*)
emigration (*e*)
biotic potential
exponential population growth
environmental resistance
carrying capacity (*K)*
r selection
r strategists (or *r*-selected species)
K selection
K strategists (or *K*-selected species)
survivorship
density-dependent factor
density-independent factor
zero population growth
demographics
highly developed countries (or developed countries)
infant mortality rates
developing countries
moderately developed countries
less developed countries (LDCs)
doubling time
replacement-level fertility
total fertility rate
demographic transition
preindustrial stage
transitional stage
industrial stage
postindustrial stage
age structure
age structure diagram
population growth momentum
Baby Boom
Immigration and Nationality Act
Immigration Reform and Control Act (IRCA)

Multiple Choice:

1. The growth rate (*r*) of a population is determined by using the
   a. birth rate
   b. population density
   c. death rate
   d. a. and b.
   e. a. and c.

2. A shrinking population may have a _____ death rate, a _____ birth rate, and/or a _____ rate of emigration.
   a. high, low, low
   b. high, low, high
   c. high, high, low
   d. low, low, high
   e. low, high, high

3. Which of the following does *not* help determine a species' biotic potential?
   a. age at which reproduction begins
   b. number of reproductive periods per lifetime
   c. rates of immigration and emigration among populations
   d. life history characteristics
   e. number of offspring produced during each reproductive period

4. The growth curve in this figure (Figure 8-3b in *Environment*, 3/e) is the result of a(n) _____ reproductive rate; the graph illustrates _____ growth.
   a. accelerating, logistic
   b. accelerating, exponential
   c. constant, logistic
   d. constant, exponential

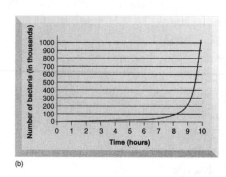

5. As a population increases in size, environmental resistance _____ and _____ population growth. Such an effect is an example of _____ feedback.
   a. increases, limits, positive
   b. increases, limits, negative
   c. decreases, promotes, positive
   d. decreases, promotes, negative
   e. increases, promotes, negative

6. Populations _____ stabilize at the environment's carrying capacity (*K*), which _____ over time.
   a. normally, changes
   b. normally, remains fixed
   c. rarely, changes
   d. rarely, remains fixed
   e. never, changes

7. A population of reindeer introduced to one of Alaska's Pribilof Islands in 1910 experienced exponential growth for about 25 years. Then the population began a steep decline, ultimately dropping from about 2000 reindeer to only 8. The reason for this decline was
   a. starvation
   b. predation
   c. overhunting
   d. infectious disease
   e. parasitism

8. Which of the following traits is *not* typical of *r* strategists?
   a. large body size
   b. short life span
   c. large broods
   d. little or no parental care of the young
   e. high population growth rate

9. Which of the following traits is *not* typical of *K* strategists?
   a. late reproduction
   b. long life span
   c. populations close to environment's carrying capacity
   d. most common in unstable environments
   e. parental care of the young

10. This figure (Figure 8-7 in *Environment*, 3/e) shows three survivorship curves. Type III most closely matches that of
    a. herring gulls
    b. oysters
    c. humans
    d. lizards
    e. reindeer

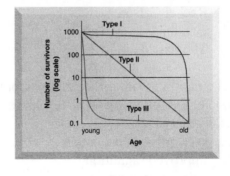

11. *K* strategists would be most likely to fit which of the survivorship curves shown above?
    a. Type I
    b. Type II
    c. Type III

12. Which of the following is *not* a density-dependent factor?
    a. predation
    b. competition
    c. disease
    d. severe weather

13. Density-dependent factors tend to regulate a population at a(n) _____ size that is _____ the environment's carrying capacity.
    a. unpredictable, below
    b. unpredictable, above
    c. constant, well below
    d. constant, near
    e. constant, above

14. Population sizes of arctic mosquitoes are determined almost entirely by a density- _____ factor, namely _____.
    a. dependent, severe winter weather
    b. independent, severe winter weather
    c. dependent, food availability
    d. independent, food availability
    e. dependent, predation

15. One of the first people to recognize that the human population cannot increase indefinitely was
    a. Herbert Spencer
    b. Charles Darwin
    c. Rachel Carson
    d. Gandhi
    e. Thomas Malthus

16. The rapid growth of the human population in the past 200 years is the result of all but one of the following. Which is *not* a cause of human population increase in the past 200 years?
    a. greater food production
    b. improved sanitation
    c. a dropping death rate
    d. improved medical care
    e. a rising birth rate

17. Zero population growth is expected to occur
    a. early in the 21st century
    b. near the middle of the 21st century
    c. near the end of the 21st century
    d. in the first half of the 22nd century
    e. in the second half of the 22nd century

18. The International Institute for Applied Systems Analysis (IIASA) predicts that the world's human population will peak at _____ billion.
    a. 7
    b. 9
    c. 11
    d. 13
    e. 15

19. The average number of children born to each woman on Earth in the year 2000 was
    a. 2.0
    b. 2.5
    c. 3.0
    d. 3.5
    e. 4.0

20. Which of the following is *not* characteristic of a less developed country?
    a. high birth rate
    b. high infant mortality rate
    c. low average per-capita GNP
    d. long doubling time
    e. short life expectancies

21. Which demographic stage is characterized by a declining birth rate and slowing population growth?
    a. preindustrial
    b. transitional
    c. industrial
    d. postindustrial
    e. all of the above

22. Which demographic stage is characterized by high birth and death rates?
    a. preindustrial
    b. transitional
    c. industrial
    d. postindustrial
    e. none of the above

23. Which demographic stage is characterized by the most rapid population growth?
    a. preindustrial
    b. transitional
    c. industrial
    d. postindustrial

24. An age-structure diagram shaped like a pyramid represents a population that is
    a. stable
    b. growing slowly
    c. shrinking slowly
    d. growing rapidly
    e. shrinking rapidly

25. Many countries in _____ have stable populations.
    a. South America
    b. sub-Saharan Africa
    c. Asia
    d. Europe
    e. northern Africa

26. In the year 2000, _____% of the world's population lived in developing countries.
    a. 40
    b. 50
    c. 60
    d. 70
    e. 80

27. One of the highest rates of population growth in a developed country is found in
    a. the U.S.
    b. Japan
    c. France
    d. Denmark
    e. Russia

28. U.S. immigration policy relaxed during World War II because
    a. of labor shortages
    b. so many men died in the fighting that U.S population began to shrink
    c. the nation's attention was focused on other issues
    d. illegal immigrants were recruited as soldiers
    e. all of the above

29. Four out of every five legal immigrants come to the U.S. from
    a. Europe or Africa
    b. Latin America or Asia
    c. Asia or Africa
    d. Europe or Asia
    e. Africa or Latin America

30. Based on data from the year 2000, the U.S. is the _____ most populous country in the world.
    a. second
    b. third
    c. fourth
    d. fifth
    e. sixth

Matching: Match the phrases on the left with the responses on the right. Some responses are used more than once.

_____ 1. a less developed country

_____ 2. had an annual increase of 0.6% in 2000

_____ 3. home of most of the world's HIV-infected children

_____ 4. has a declining population

_____ 5. a country in the postindustrial state that has a growing population

_____ 6. began to limit immigration in 1875

_____ 7. a moderately developed country

_____ 8. experienced at least a 25% drop in its fertility rate between 1990 and 2000

_____ 9. has an age-structure diagram that is tapered ("pinched") at the bottom

_____ 10. is experiencing declining life expectancies in some areas

a. sub-Saharan Africa

b. Germany

c. Mexico

d. Bangladesh

e. the U.S.

Fill-In:

1. _____ (two words) is a discipline that deals with the number of individuals of a species found in an area and how and why that number changes over time.

2. _____ (two words) is the number of individuals of a species per unit area or volume at a given time.

3. In human populations, growth rate is also referred to as _____ (two words).

4. _____ involves movement of individuals into a population and a consequent increase in that population's size.

5. A population's _____ (two words) is the maximum rate at which it could increase under ideal conditions.

6. Factors called _____ (three words) determine the biotic potential of a species.

7. The limits the environment sets to population growth are collectively called _____ (two words).

8. The largest population that can theoretically be supported, indefinitely, by a particular environment is called the _____ (two words).

9. Species with traits that contribute to a high population growth rate are called _____ strategists.

10. Tawny owls and redwood trees are examples of _____ strategists, species that compete well in stable environments.

11. _____ is the proportion of individuals in a population that survive to a particular age.

12. If a change in population density affects an environmental factor's impact on a population, that factor is said to be density-_____.

13. Abiotic factors such as random weather events are generally density-_____ factors.

14. The worldwide growth rate (*r*) peaked at _____% per year in 1962–63.

15. The main unknown factor in the U.N.'s (and other organizations') population growth scenario is Earth's _____ (two words).

16. _____ is the applied branch of sociology that deals with human population statistics.

17. _____ fertility is the number of children a couple must have in order to "replace" themselves.

18. The worldwide _____ (two words) rate—the average number of children born to each woman—is 3.0.

19. A population's _____ (two words) is the number and proportion of people at each age.

20. Worldwide, _____% of the human population is under age 15.

21. The _____ (four words) Act (IRCA) is the basic immigration law of the U.S.

22. The state with the highest percentage of people under the age of 18 is _____; the state with the highest percentage of senior citizens (over age 65) is _____. In 2020, the three states with the largest human populations are predicted to be _____, _____, and _____.

Short Answer:

1. What factors contribute to the high incidence of HIV infection in southern Africa?

2. What four factors determine the growth rate (*r*) of a population?

1. What aspects of a population's environment determine environmental resistance? How does environmental resistance change as a population becomes crowded?

2. Explain how nature requires that organisms make tradeoffs in their expenditure of energy.

3. Explain how tawny owls regulate their population size. Are they *r* or *K* strategists?

4. Explain why herring gulls do not fit a single survivorship curve.

5. On what assumptions do projections of human population growth depend?

6. Why is replacement-level fertility always greater than 2.0? Why do developing countries have higher replacement levels than highly developed countries do?

7. What factors contribute to population stabilization in highly developed countries?

8. What are the two reasons for continuing population growth in the U.S.?

9. What are some of the threats and obstacles faced by refugees?

Critical-Thinking Questions:

1. Blue jays mature rapidly and are likely to reproduce the year after they are born. Blue jay parents invest heavily in the care of their young, and an adult is likely to

reproduce more than once. Would you classify the blue jay as an *r*-selected or a *K*-selected species? Justify your answer.

2. The U.S. abolished national immigration quotas in 1965. Do you see the priority system, legislated by the revised Immigration Reform and Control Act (IRCA), as an improvement? Why or why not? How do you feel about an "open door" policy that does not attempt to limit immigration? Justify your support or rejection of such a policy.

3. Describe the dynamic relationship between wolves and moose on Lake Superior's Isle Royale. If the wolves die out, how do you think the moose population will change? Overall, do you see the wolves as a benefit or a detriment to the forest communities of the island? Explain.

Data Interpretation:

1. Suppose a population of 100,000 has 1200 births (12 per 1000), 600 deaths (6 per 1000), 150 immigrants (1.5 per 1000), and 80 emigrants (0.8 per 1000) per year. What is its rate of population growth?

2. Suppose a population of 10,000 has 80 births, 40 deaths, 7 emigrants, and a growth rate of 0.009 per year. What is its immigration rate?

3. If human population continues to grow at the rate of 1 billion every 12 years, what will its size be in the year 2029? (Start with the 1999 population size.)

Chapter 9

Facing the Problems of Overpopulation

Study Outline:

I. The Human Population Explosion
    A. Population and World Hunger
    B. Economic Effects of Continued Population Growth

II. Population, Resources, and the Environment
    A. Types of Resources
    B. Population Size and Resource Consumption
    C. People Overpopulation and Consumption Overpopulation
    D. Population and Environmental Impact: A Simple Model

III. Population and Urbanization
    A. Characteristics of the Urban Population
    B. The City as an Ecosystem
    C. Environmental Problems Associated with Urban Areas
    D. Environmental Benefits of Urbanization

IV. Case in Point: Curitiba, Brazil
    A. Urbanization Trends

V. Reducing the Total Fertility Rate
    A. Culture and Fertility
    B. The Social and Economic Status of Women
    C. Family Planning Services

VI. Government Policies and Fertility

VII. Case in Point: China's Controversial Family Planning Policy

VIII. Case in Point: India's Severe Population Pressures

IX. Case in Point: Mexico's Young Age Structure

Key Terms: The following terms are listed in order of appearance in your textbook.

total fertility rate (TFR)
nonrenewable resources
renewable resources
consumption
people overpopulation
consumption overpopulation
urbanization
urban heat island
compact development
pronatalists

Multiple Choice:

1. Approximately _____% of the people in the world live in less developed (or "developing") countries.
   a. 40
   b. 50
   c. 60
   d. 70
   e. 80

2. According to demographers at the Population Institute, _____ is the most urgent global population problem.
   a. rising birth rates
   b. rapid growth in the poorest countries
   c. rapid growth in the wealthiest countries
   d. overconsumption
   e. rising death rates

3. The two regions of the world with the greatest food insecurity are _____ and _____.
   a. South Asia, South America
   b. China, Western Asia
   c. sub-Saharan Africa, South Asia
   d. South America, sub-Saharan Africa
   e. Central America, North Africa

4. Which of the following is *not* a renewable resource?
   a. forests
   b. fertile soil
   c. fresh water
   d. natural gas
   e. fisheries

5. During the 1990s, the average fuel economy of new cars in the U.S. _____ because of the popularity of _____.
   a. increased, compact cars
   b. decreased, minivans
   c. decreased, sport utility vehicles (SUVs)
   d. decreased, mid-sized sedans
   e. increased, mid-sized sedans

6. Highly developed countries represent _____% of the world's population and consume _____ of its resources.
   a. 50, about half
   b. 50, more than half
   c. 30, less than half
   d. 20, about half
   e. 20, more than half

7. Highly developed nations generate _____% of the world's pollution and waste.
   a. 35
   b. 45
   c. 55
   d. 65
   e. 75

8. In the U.S. today, approximately _____% of the population is involved in farming, while _____% of the population lives in cities.
    a. 40, 45
    b. 25, 55
    c. 15, 65
    d. 5, 75
    e. 2, 90

9. Compared to the population of the surrounding countryside, people in urban areas tend to be _____ and more _____.
    a. younger, heterogeneous
    b. older, heterogeneous
    c. younger, homogeneous
    d. older, homogeneous

10. A person living in a city experiences _____ temperatures and _____ precipitation than a person in the surrounding countryside does.
    a. warmer, more
    b. warmer, less
    c. cooler, more
    d. cooler, less

11. Curitiba, Brazil, serves as an example of
    a. urban sprawl
    b. compact development
    c. severe urban air pollution
    d. consumption overpopulation
    e. people overpopulation

12. Mexico City serves as an example of
    a. compact development
    b. consumption overpopulation
    c. the success of family planning services
    d. rapid, uncontrolled urban growth
    e. minimal urban air pollution

13. High total fertility rates are associated with
    a. high infant and child mortality rates
    b. low infant and child mortality rates
    c. consumption overpopulation
    d. high education levels of women
    e. high social status of women

14. A society in which children contribute to the family's livelihood and later support
    their aging parents is likely to have a
    a. high total fertility rate
    b. low total fertility rate
    c. high rate of consumption
    d. low rate of consumption
    e. low infant mortality rate

15. In most _____ countries, a higher percentage of _____ than _____ are illiterate.
    a. developing, men, women
    b. developing, women, men
    c. highly developed, men, women
    d. highly developed, women, men

16. Growing evidence indicates the single most important factor contributing to high total
    fertility rates is
    a. low survival rates of infants
    b. a tradition of large family size
    c. low status of women
    d. religious beliefs discouraging contraception
    e. urbanization

17. In the U.S., women who _____ have the highest fertility rates.
    a. complete graduate school
    b. complete college
    c. complete high school
    d. do not complete high school

18. When used correctly, _____ is/are the most effective form of birth control.
    a. condoms
    b. IUDs
    c. oral contraceptives
    d. spermicides
    e. Depo-Provera

19. The world's most rapid and significant decline in fertility occurred in _____ between
    1970 and 1981.
    a. Mexico
    b. the U.S.
    c. India
    d. Nigeria
    e. China

20. Which of the following countries has had the least success in lowering its total fertility rate?
    a. Mexico
    b. the U.S.
    c. India
    d. Nigeria
    e. China

21. Which of the following countries is expected to have a vast excess of marriageable males by the middle of the 21st century?
    a. Mexico
    b. the U.S.
    c. India
    d. Nigeria
    e. China

22. Which of the following countries, in the 1970s, used compulsory sterilization of males as a way to slow population growth?
    a. Mexico
    b. the U.S.
    c. India
    d. Nigeria
    e. China

23. Both Egypt and Mexico use _____ to reduce total fertility rates.
    a. religious doctrines
    b. soap operas with family planning messages
    c. compulsory sterilization
    d. severe economic penalties
    e. high employment rates

24. In _____, nearly half the population is below age 15, and half of all women are married by age 17.
    a. Mexico
    b. the U.S.
    c. India
    d. Nigeria
    e. China

25. *Voluntary simplicity* is a term that applies to efforts to reduce
    a. consumption
    b. family income
    c. family size
    d. urbanization
    e. education

26. Recently, the National Science Foundation added _____ to its list of long-term ecological research (LTER) sites.
    a. deserts
    b. lakes
    c. wetlands
    d. suburbs
    e. cities

27. Maternal deaths from botched illegal abortions are most common in
    a. Latin America
    b. Japan
    c. Russia
    d. India
    e. China

28. Most pronatalists are found in
    a. Latin America
    b. the U.S.
    c. Europe
    d. India
    e. Southeast Asia

29. In 2000, the world's largest city was
    a. New York, USA
    b. Tokyo, Japan
    c. Mexico City, Mexico
    d. Calcutta, India
    e. Bombay, India

Matching: Match the terms on the left with the responses on the right.

_____ 1. total fertility rate (TFR)
_____ 2. nonrenewable resources
_____ 3. renewable resources
_____ 4. consumption
_____ 5. National Research Council
_____ 6. Worldwatch Institute
_____ 7. urbanization
_____ 8. 1994 Global Summit on Population and Development
_____ 9. U.N. Population Fund
_____ 10. pronatalists

a. minerals and fossil fuels
b. calculates the quantity of materials consumed annually in the U.S.
c. supports international family-planning programs
d. promote population growth
e. the average number of children born to each woman
f. the human use of energy and resources
g. decreases total fertility rates
h. fresh water, forests, soils, and fisheries
i. examined knowledge about consumption in the U.S.
j. placed particular emphasis on the empowerment of women

Fill-In:

1. The fastest rates of population growth are found in _____ (developed or developing) countries.

2. In order to feed its projected population in 2050, Africa will have to increase its food production by _____%.

3. In 1992, the U.S. National Academy of Sciences and the Royal Society of London forecast irreversible _____ (two words) and continued _____ as results of ongoing population growth.

4. In order for a country's standard of living to rise, the country's _____ growth must be greater than its _____ growth.

5. Renewable resources are only *potentially* renewable, meaning they must be used in _____ ways.

6. The economic growth of developing countries is often tied to the exploitation of _____ (two words).

7. A country is _____ if demand on its resources damages the environment.

8. _____ overpopulation occurs when each member of a population consumes too large a share of resources; _____ overpopulation occurs when a population's size causes environmental deterioration.

9. Cities in developing countries tend to have _____ (more or fewer) males than females.

10. Urban ecologists use the acronym POET to refer to the variables that shape the urban "ecosystem." P = _____, O = _____, E = _____, and T = _____.

11. Abandoned areas in cities are called _____.

12. The worst air pollution in the world is found in _____ (rural or urban) areas in _____ (developing or highly developed) nations.

13. _____ (two words) is an urban planning strategy that relies heavily on public transportation and close proximity of housing to jobs and shopping sites.

14. Most of the world's largest cities are in _____ (developing or highly developed) countries.

15. Urbanization is occurring at a _____ (faster or slower) rate in developing nations than it is in developed nations.

16. There is always a correlation between marriage age and _____ (three words).

17. _____ (Larger or Smaller) family size is associated with increased family income.

18. The U.N. Fund for _____ (two words) is an important source of money for family planning projects in developing countries.

19. China's mid-2000 population was estimated to be _____ people; its total fertility rate of _____ was below the replacement rate.

20. The migration of Mexicans to the U.S. is primarily due to _____ (two words) in Mexico.

21. Reducing infant and child mortality rates through immunization results in _____ (higher or lower) birth rates.

22. As developing countries go, Brazil is unusual in that it has a _____ (higher or lower) illiteracy rate for young men than young women.

Short Answer:

1. What three approaches to the world hunger problem are presented in the text?

2. How does debt interfere with efforts to raise standards of living in developing countries?

3. What two generalizations are made about the link between population growth and natural resources?

4. How is poverty linked in a vicious cycle with environmental degradation?

5. What is the equation that is used to assess human impact on the environment?

6. Why have urban populations grown at the expense of rural populations?

7. Name some of the environmental problems and benefits of urbanization.

8. What are the particular challenges of cities in developing nations?

9. The percentage of women using family-planning services to limit family size increased from less than 10% in the 1960s to more that 50% during the 1990s. Has the actual number of women *not* using family-planning services decreased? Explain.

10. In what ways do government-enforced policies affect fertility rates?

11. What strategies have been used in recent years to slow India's population growth?

12. Underfunding of the World Programme of Action is impeding efforts to slow population growth. What are some of the direct results of underfunding predicted by the U.N. Population Fund?

Critical-Thinking Questions:

1. Why is it that developed countries, such as the U.S., which make a proportionately small contribution to world population, affect the environment as much or more than the population explosion in developing countries? What steps might alleviate this pressure on the environment?

2. Compare the city you live in, or the city closest to where you live, to Curitiba, Brazil. Are there ways in which your city could be redesigned to improve both quality of life and the quality of environment?

3. Do you believe governments are justified in trying to lower (or, in some cases, raise) fertility rates? What population policy would you support in your own country? Justify your responses.

Data Interpretation:

1. Determine the environmental impact, in kilograms of particulate pollution per year, of 150 million Americans driving 13,000 miles per year and releasing .03 g of particulates per mile.

2. Suppose 40% of women in a developing country have a secondary or higher education. These women have an average of 2.8 children each. Suppose 40% of women in this country have a primary education and have an average of 4.7 children each, while 20% of women have no formal education and have an average of 5.5 children each. What is the total fertility rate for this country?

3. China's total fertility rate dropped from 5.8 in 1970 to 1.8 in 2000. What percent of the 1970 fertility rate is represented by the 2000 fertility rate?

PART FOUR:

THE SEARCH FOR ENERGY

Chapter 10

Fossil Fuels

Study Outline:

Key Terms: The following terms are listed in order of appearance in your textbook.

Strategic Petroleum Reserve
fossil fuel
nonrenewable resources
hydrocarbons
methane
lignite
bituminous coal (or soft coal)
sub-bituminous coal
anthracite (or hard coal)
surface mining
subsurface mining
black lung disease
Surface Mining Control and Reclamation Act (SMCRA)
acid mine drainage
dragline
acid deposition
acid precipitation
scrubbers
resource recovery
Clean Air Act Amendments of 1990
clean coal technologies
fluidized-bed combustion
petroleum (or crude oil)
petrochemicals
liquefied petroleum gas
cogeneration
structural traps
anticline
strata
salt domes
continental shelves

Oil Pollution Act
synthetic fuels (or synfuels)
tar sands (or oil sands)
oil shales
gas hydrates (or methane hydrates)
coal liquefaction
coal gasification
Energy Policy Act of 1992
subsidy

Multiple Choice:

1. The speed limit was lowered to 55 mph in the 1970s
    a. to save human lives
    b. as an energy conservation measure
    c. to reduce air pollution
    d. to reduce damage to aging bridges
    e. to reduce the number of wildlife roadkills

2. During the late 1980s, U.S. oil imports _____ dramatically, while domestic oil production _____.
    a. decreased, increased
    b. decreased, remained unchanged
    c. increased, increased
    d. increased, decreased
    e. increased, remained unchanged

3. Most of the energy used in North America is supplied by
    a. fossil fuels
    b. nuclear fission
    c. solar power
    d. hydroelectric power
    e. wood and charcoal

4. _____ comes from the heated and compressed remains of ancient swamp plants.
    a. Sedimentary rock
    b. Oil
    c. Methane
    d. Propane
    e. Coal

5. Methane is the most common form of
   a. oil
   b. coal
   c. natural gas
   d. fossil fuel
   e. lignite

6. In the 18th century, _____ began to be replaced by _____ as the dominant fuel.
   a. wood, coal
   b. oil, coal
   c. coal, oil
   d. wood, natural gas
   e. coal, natural gas

7. The most common type of coal, which often contains sulfur, is
   a. anthracite
   b. bituminous coal
   c. lignite
   d. sub-bituminous coal
   e. hard coal

8. A soft coal that produces less heat than other coals is
   a. anthracite
   b. bituminous coal
   c. lignite
   d. sub-bituminous coal
   e. hard coal

9. You have in your hand a lump of coal, said to come from Pennsylvania, that is a deep, shiny black. You recognize it as
   a. anthracite
   b. bituminous coal
   c. lignite
   d. sub-bituminous coal
   e. soft coal

10. The cleanest and hottest burning coal is
    a. anthracite
    b. bituminous coal
    c. lignite
    d. sub-bituminous coal
    e. soft coal

11. The low-sulfur coal burned by many coal-fired electric power plants in the United States is
    a. anthracite
    b. bituminous coal
    c. lignite
    d. sub-bituminous coal
    e. soft coal

12. The most abundant fossil fuel in the world is
    a. oil
    b. natural gas
    c. gasoline
    d. coal
    e. methane

13. Coal miners run the risk of developing a pulmonary disease called
    a. pleurisy
    b. pneumonia
    c. emphysema
    d. tuberculosis
    e. black lung disease

14. Erosion, water pollution, landslides, and loss of wildlife habitat are problems associated with extraction of
    a. coal
    b. oil
    c. natural gas
    d. methane
    e. petrochemicals

15. Severe water pollution is a risk associated with the _____ industries.
    a. coal and natural gas
    b. oil and natural gas
    c. coal and oil
    d. oil only
    e. coal only

16. Acid deposition is caused largely by the burning of _____, which may release sulfur dioxide as well as oxides of nitrogen.
    a. wood
    b. oil
    c. natural gas
    d. coal
    e. petrochemicals

17. Resource recovery in the coal industry
    a. reduces the use of coal
    b. reduces pollution
    c. creates a marketable product from wastes
    d. a and c
    e. b and c

18. Enforcement of the Clean Air Act has greatly reduced emissions of _____ from coal-burning power plants.
    a. carbon dioxide
    b. butane
    c. sulfur dioxide
    d. lead
    e. nitrous oxides

19. In the 1940s, _____ began to replace _____ as the most important energy source(s) in the United States.
    a. coal, wood
    b. coal, oil and natural gas
    c. oil and natural gas, wood
    d. oil and natural gas, coal
    e. hydroelectric power, wood

20. Petrochemicals, used in the production of fertilizers, plastics, paints, pesticides, medicines, and synthetic fibers, are derived from
    a. bituminous coal
    b. crude oil
    c. liquefied petroleum gas
    d. lignite
    e. methane

21. More than half of the world's oil reserves are located in
    a. Russia
    b. Persian Gulf countries
    c. the United States
    d. Libya
    e. China

22. The U.S. imports
    a. more than one-half of its oil
    b. less than one-half of its oil
    c. nearly all of its coal
    d. more than one-half of its natural gas
    e. nearly all of its natural gas

23. Gasoline combustion releases _____ into the atmosphere, thereby contributing to
    _____.
    a. nitrogen oxides, acid deposition
    b. carbon dioxide, global warming
    c. sulfur dioxide, photochemical smog
    d. b and c
    e. a and b

24. The advantage of burning natural gas instead of oil is that natural gas
    a. is cheaper
    b. is more available
    c. pollutes much less
    d. yields better gas mileage than gasoline
    e. produces more heat

25. The Oil Pollution Act of 1990 requires that all oil tankers in U.S. waters _____ by
    2015.
    a. be outlawed
    b. have double hulls
    c. carry accident insurance
    d. use radar
    e. have environmental inspectors on board

26. A 1995 study by the Department of the Interior concluded that _____ in the Arctic
    National Wildlife Refuge _____ harm the environment.
    a. oil drilling, would
    b. oil drilling, would not
    c. coal mining, would
    d. coal mining, would not
    e. oil-shale mining, would not

27. _____, once the major fuel for lighting and heating American homes, burn(s) almost
    as cleanly as natural gas.
    a. Oil shales
    b. Tar sands
    c. Gas hydrates
    d. Coal gas
    e. Liquefied coal

28. Vast areas of land would have to be surface-mined to exploit _____ and _____.
    a. gas hydrates, oil shales
    b. tar sands, oil shales
    c. gas hydrates, tar sands
    d. methane hydrates, oil shales
    e. liquefied coal, methane hydrates

29. A car uses approximately _____% more gasoline at 75 mph than it does at 55 mph.
    a. 10
    b. 20
    c. 30
    d. 40
    e. 50

30. The cost of gasoline is much _____ in Japan and Europe than it is in the United
    States; more expensive gasoline _____ seem to hamper a nation's economic
    competitiveness.
    a. lower, does
    b. lower, does not
    c. higher, does
    d. higher, does not

Matching: Match the terms on the left with the responses on the right.

_____ 1. hydrocarbons   a. burns a coal-limestone mixture; reduces pollution
_____ 2. highwalls   b. a synfuel abundant in Venezuela and Alberta, Canada
_____ 3. scrubbers   c. using natural gas to produce electricity and generate
_____ 4. resource recovery   steam for water and space heating
_____ 5. fluidized-bed   d. cliffs of excavated rock at a surface mine
         combustion   e. desulfurization systems; reduce coal's sulfur emissions
_____ 6. petroleum   f. a liquid composed of many hydrocarbon compounds
_____ 7. liquefied petroleum   g. used for home lighting and heating until replaced by oil
         gas   and natural gas
_____ 8. cogeneration   h. molecules that make up oil
_____ 9. anticline   i. underground, ice-encrusted natural gas
_____10. salt dome   j. rock strata, folded upward, that may trap oil or natural
_____11. tar sands   gas
_____12. oil shales   k. creates a marketable product from industrial waste
_____13. gas hydrates   l. used mainly for heating and cooking in rural areas
_____14. coal gas   m. an underground column of salt that may trap oil or
                                  natural gas
                               n. a synfuel that is not yet cost-efficient to utilize

Fill-In:

1. The U.S. energy crisis of 1973 was precipitated by OPEC's decision to restrict
    _____ shipments to the U.S.

2. What does "OPEC" stand for? _____

3. An important part of U.S. energy security policy is the
   _____ (three words), an emergency supply of oil.

4. Per capita energy use in developed countries is approximately _____ times greater
   than in developing countries.

5. Deposits of _____ and _____ (two types of
   fossil fuel) are often found together.

6. _____ (type of fossil fuel) supplied the energy that fueled the
   Industrial Revolution.

7. Today, coal is used primarily to produce _____.

8. The grade of coal common to the Appalachian, Great Lakes, and Mississippi regions
   is _____.

9. The highest grade of coal is _____.

10. The United States has massive deposits of _____ (type of fossil
    fuel), representing 23.6% of the world's supply.

11. A _____ is a $100-million machine that removes entire
    mountaintops to extract coal.

12. Burning *any* fossil fuel releases _____.

13. The burning of _____ (type of fossil fuel) releases the most carbon
    dioxide per unit of heat produced.

14. Oxides of sulfur and nitrogen form _____ when they react with
    water in the atmosphere.

15. The normal pH of rain is _____, which is slightly _____ (basic or
    acidic).

16. _____ (type of fossil fuel) is the source of gasoline, diesel oil,
    asphalt, plastics, and many other products.

17. _____ (two words) contains methane, ethane, propane, and
    butane.

18. It is believed that large oil deposits exist under and adjacent to
    _____ (two words).

19. _____ has more reserves of natural gas than any other country.

20. What U.S. city has the most buses powered by natural gas? _____

21. Energy experts predict that the United States will be importing nearly 100% of its _____ (type of fossil fuel) by 2015.

22. A study conducted by the Department of the Interior concluded there is a _____% chance of finding oil in the Arctic National Wildlife Refuge.

23. The Arctic tundra has massive deposits of _____.

24. By 2025, _____ (what country) may be releasing more than one-half the amount of carbon dioxide currently released worldwide.

25. Because of government _____, gasoline prices in the United States _____ (do/do not) reflect the true cost of gasoline.

Short Answer:

1. What were some effects in the United States of the oil embargo of 1973?

2. What has caused increasing gasoline consumption in the United States in the past two decades?

3. Refer to Figure 10-2 in *Environment*, 3/e, to list the two biggest uses of energy in the United States.

4. What is SMCRA, and what limits its effectiveness?

5. In what ways do surface mines cause water pollution?

6. How do carbon dioxide emissions contribute to global warming, and what are some of the effects of global warming?

7. What advantages of burning oil and natural gas led these fossil fuels to upstage coal in the 1940s?

8. Why is it difficult to predict how long oil and natural gas supplies will last?

9. In what ways is natural gas superior to other fossil fuels?

10. What types of organisms were killed by the 1989 Alaskan oil spill?

11. What arguments are made for and against oil exploration in the Arctic National Wildlife Refuge?

12. What are the four objectives of the U.S. energy strategy?

13. Name several things we can do to decrease our use of gasoline.

Critical-Thinking Questions:

1. Take a look at the uses of energy mentioned at the beginning of the chapter. Do you feel that a high standard of living is *necessarily* linked to prodigious energy consumption? How might we reduce energy use, while maintaining a high standard of living?

2. The text gives estimates of the longevity of known coal deposits based on *the present rate of consumption*. Why is the present rate of consumption an unreliable measure of the future? How do you think consumption of coal will change? Justify your answer.

3. In what ways does the United States make itself vulnerable by relying heavily on imported oil?

4. Some U.S. energy experts say we should "drain America first"; others say we should "deplete foreign oil reserves… and save domestic supplies for the future." Which approach do you favor and why? Or, if you reject both, suggest and support a better alternative.

5. See the photograph (with "Meeting the Challenge" essay) of reclaimed coal-mined land. The caption says, "…the land is similar to what it was before coal mining occurred." In what way(s) is it different? Do you think the land in Figure 10-6 in *Environment*, 3/e will ever be restored to its former appearance and species composition? Justify your answer.

Data Interpretation:

1. The heat value in BTU/pound of anthracite is _____ % that of lignite. (See Table 10-1 in *Environment*, 3/e.)

2. See Table 10-3 in *Environment*, 3/e. How many times more BTUs of energy is used per person per mile when someone drives a car instead of taking a bus?

3. If gasoline trades for $2.00 per gallon on the world market but sells for $1.30 per gallon in the United States, the U.S. government subsidizes energy at a cost of _____ per gallon.

4. A car driven at 75 mph consumes approximately 50% more fuel than when driven at 55 mph. Suppose your car gets 36 miles per gallon at 55 mph; what will its fuel economy be at 75 mph?

Chapter 11

Nuclear Energy

Study Outline:

I.   Introduction to Nuclear Processes
        A. Atoms and Radioactivity

II.   Conventional Nuclear Fission
        A. How Electricity Is Produced from Nuclear Energy
        B. Safety Features of Nuclear Power Plants

III.   Breeder Nuclear Fission

IV.   Is Nuclear Energy a Cleaner Alternative than Coal?

V.   Is Electricity Produced by Nuclear Energy Cheap?
        A. The High Cost of Building a Nuclear Power Plant
        B. The High Cost of Fixing Technical and Safety Problems in Existing Plants

VI.   Can Nuclear Energy Decrease Our Reliance on Foreign Oil?

VII.   Safety in Nuclear Power Plants

VIII.   Case in Point: Three Mile Island

IX.   Case in Point: Chernobyl

X.   The Link Between Nuclear Energy and Nuclear Weapons

XI.   Radioactive Wastes

XII.   Case in Point: Yucca Mountain
        A. High-Level Liquid Waste
        B. Radioactive Wastes with Relatively Short Half-Lives
        C. Decommissioning Nuclear Power Plants

Key Terms: The following terms are listed in order of appearance in your textbook.

fission
fusion
atomic mass
atomic number
isotopes
deuterium
tritium
radioisotopes
radioactive
radiation
radioactive decay
radioactive half-life
nuclear fuel cycle
enrichment
nuclear reactor
fuel rods
fuel assemblies
reactor core
steam generator
turbine
condenser
control rod
primary water circuit
secondary water circuit
tertiary water circuit
cooling tower
reactor vessel
containment building
breeder nuclear fission

spent fuel
meltdown
low-level radioactive wastes
Low-Level Radioactive Policy Act
high-level radioactive wastes
Nuclear Waste Policy Act
Vitrification
storage
entombment
decommission
plasma
ionizing radiation
mutations
oncogenes

Multiple Choice:

1. Nuclear reactions involve
    a. changes in the chemical bonds between atoms
    b. changes in the nuclei of atoms
    c. the conversion of matter into energy
    d. a and c
    e. b and c

2. Isotopes of hydrogen differ in their number of
    a. protons
    b. neutrons
    c. electrons
    d. nuclei
    e. a and b

3. An element's atomic number is determined by the number of _____ in each atom.
    a. protons
    b. neutrons
    c. electrons
    d. nuclei
    e. a and b

4. Deuterium and tritium are isotopes of
    a. hydrogen
    b. helium
    c. uranium
    d. plutonium
    e. carbon

5. The continent with the largest deposits of uranium ore is
    a. Africa
    b. South America
    c. Europe
    d. Australia
    e. North America

6. In nuclear power plants, _____ is generated, which is used to produce _____, which is used to drive turbines, which generate _____.
    a. electricity, heat, steam
    b. steam, electricity, heat
    c. heat, steam, electricity
    d. heat, electricity, steam
    e. electricity, steam, heat

7. In *conventional* nuclear fission, the fuel is _____, while in *breeder* nuclear fission, the fuel may be _____.
    a. strontium-90, cerium-144
    b. uranium-235, plutonium-239
    c. cesium-137, krypton-85
    d. xenon-133, radon-226
    e. plutonium-239, cerium-144

8. A control rod controls the rate of fission by absorbing
    a. heat
    b. electrons
    c. neutrons
    d. water
    e. protons

9. Nuclear power plants have large cooling towers that cool
    a. spent fuel rods
    b. the reactor core
    c. uranium-238
    d. control rods
    e. water in the tertiary water circuit

10. A containment building, with 3- to 5-foot-thick walls, serves as
    a. a storage site for radioactive wastes
    b. a storage site for uranium ore
    c. a protective defense against radioactive leaks from a nuclear reactor
    d. the headquarters of a nuclear power plant
    e. an emergency shelter in the event of a nuclear accident

11. Approximately _____ of the electricity generated in the United States comes from nuclear power plants.
    a. 1/5
    b. 1/3
    c. 1/2
    d. 2/3
    e. 4/5

12. The use of nuclear energy cannot reduce our nation's use of _____ very much because only 6 percent of our electricity is generated by burning this fuel.
    a. oil
    b. coal
    c. natural gas
    d. wood
    e. ethanol

13. In 1979, the most serious nuclear reactor accident in the United States occurred at
    a. Peach Bottom (PA)
    b. Calvert Cliffs (MD)
    c. Three Mile Island (PA)
    d. Salem (NJ)
    e. Shippingport (PA)

14. Following the Chernobyl explosion and subsequent clean-up in 1986, _____ was the strategy used on the destroyed reactor.
    a. storage
    b. entombment
    c. decommission
    d. decontamination
    e. all of the above

15. _____ is alarmingly common in children in the vicinity of the Chernobyl nuclear accident.
    a. Diabetes
    b. Cystic fibrosis
    c. Thyroid cancer
    d. Meningitis
    e. Stomach cancer

16. The region outside Eastern Europe hardest hit by radioactive fallout from the Chernobyl accident was
    a. Portugal and Spain
    b. Romania
    c. Italy
    d. Poland
    e. Scandinavia

17. According to experts at the IAEA (International Atomic Energy Agency), ten nuclear reactors in _____ pose significant safety risks.
    a. southern Europe
    b. the United Kingdom
    c. the United States
    d. Ukraine and nearby countries
    e. earthquake-prone areas

18. Ionizing radiation can lead to cancer because of the damage it does to
    a. oncogenes
    b. DNA
    c. proteins
    d. cell membranes
    e. organelles

19. Overall, the greatest source of radiation exposure to humans comes from
    a. medical X-rays
    b. the nuclear fuel cycle
    c. cosmic rays
    d. natural sources of radon
    e. consumer products

20. Which of the following in *not* an effect of high, but non-lethal, doses of radiation?
    a. hearing loss
    b. cataracts
    c. temporary male sterility
    d. skin burns
    e. leukemia

21. Storage of nuclear wastes in _____ is believed by most experts to be the safest option.
    a. above-ground mausoleums
    b. ice sheets
    c. stable rock formations
    d. the seabed
    e. containment buildings

22. In the United States, most radioactive wastes are
    a. in permanent underground storage
    b. in "temporary" on-site storage
    c. in permanent deep-sea storage
    d. no longer in need of protective storage
    e. missing and unaccounted for

23. The proposed site for a permanent storage facility for high-level nuclear wastes is Yucca Mountain in
    a. Montana
    b. Nevada
    c. Ohio
    d. Mississippi
    e. Idaho

24. The phenomena of NIMBY (Not In My BackYard) and NIMTOO (Not In My Term Of Office) interfere with
    a. the disposal of nuclear wastes
    b. nuclear energy research
    c. the building of new nuclear-power facilities
    d. the upgrading of safety features in older nuclear facilities
    e. a and c

25. The most dangerous and polluting disposal of radioactive wastes today occurs in
    a. the United States
    b. France
    c. Japan
    d. Russia
    e. China

26. High-level liquid nuclear wastes are converted into solid glass logs through a process called
    a. enrichment
    b. entombment
    c. fusion
    d. fission
    e. vitrification

27. Stars "burn" through nuclear _____; their fuel is _____.
    a. fission, hydrogen
    b. fusion, hydrogen
    c. fission, uranium
    d. fusion, uranium
    e. fusion, helium

28. Deuterium, a fuel for nuclear fusion, can be obtained from
    a. iron oxides
    b. water
    c. plutonium
    d. limestone
    e. uranium

29. A major challenge associated with fusion is
    a. disposal of high-level radioactive waste
    b. the exceptionally high temperatures required
    c. a limited fuel supply
    d. all of the above
    e. none of the above

Matching: Match the terms on the left with the responses on the right.

____ 1. fission
____ 2. fusion
____ 3. enrichment
____ 4. reactor core
____ 5. turbine
____ 6. control rod
____ 7. primary water circuit
____ 8. secondary water circuit
____ 9. tertiary water circuit
____ 10. reactor vessel
____ 11. low-level radioactive waste
____ 12. high-level radioactive waste
____ 13. entombment
____ 14. decommissioning

a. uses steam to generate electricity
b. provides cool water to the condenser, which cools steam
c. heats water by using fission energy
d. an example is waste from a nuclear weapons facility
e. involves refining uranium ore to increase the concentration of U-235
f. involves dismantling a nuclear power plant
g. involves the splitting of an atomic nucleus
h. used to regulate the rate of fission
i. an example is glassware from certain research labs
j. involves the joining of two atomic nuclei into a single nucleus
k. where fission occurs in a nuclear power plant
l. a steel, pot-like structure that surrounds the reactor core
m. converts water to steam, which turns the turbine
n. involves encasing a nuclear power plant in concrete

Fill-In:

1. Each radioactive element has its own rate of decay. This rate, defined as the time required for half of a radioactive substance to decay, is called the element's _____.

2. Uranium ore is a _____ (renewable or nonrenewable) resource.

3. Pellets of nuclear fuel are placed in closed pipes called _____ (two words).

4. In fission, energy and _____ are released from the reaction.

5. The type of fission that produces, as well as consumes, nuclear fuel is called _____ nuclear fission.

6. A nuclear accident in which high temperatures cause the metal encasing nuclear fuel to melt is called a _____.

7. According to the Ukrainian health minister, the human death toll from the Chernobyl accident was _____ lives by 1999.

8. _____ and _____ are the fuels commonly used in atomic fission weapons.

9. Changes in DNA, which can be caused by exposure to ionizing radiation, are called _____.

10. Cancer-causing genes are called _____.

11. The United States needs to dispose of more than 50 tons of plutonium from the dismantling of _____ (two words).

12. MOX, or mixed oxide, is a fuel designed to dispose of _____.

13. The _____ (three words) Act of 1980 requires states to take responsibility for the wastes they generate.

14. The Yucca Mountain site for a nuclear wastes facility is controversial because of its proximity to _____ and _____.

15. The discovery of radioactive tritium in _____ near Yucca Mountain calls into question the suitability of the site.

16. _____, _____, and _____ are radioisotopes with relatively short half-lives; their wastes are safe in 300 to 600 years.

17. Nuclear power plants are licensed to operate for no more than _____ years.

18. The first commercial nuclear power plant in the United States, at _____, Pennsylvania, was dismantled in 1989.

19. The Sun is powered by nuclear _____.

20. _____, rather than custom-building, nuclear power plants can lower costs of construction.

21. The new generation of nuclear power plants are _____ (more/less) complex and expensive to build than earlier ones were.

Short Answer:

1. What was the Manhattan Project, and what role did Albert Einstein play in it?

2. Contrast the amount of energy released in a nuclear reaction with the amount released in a chemical reaction, such as combustion.

3. Explain why radioactive wastes with a half-life of only 30 years would take hundreds of years to become safe.

4. Why do breeder nuclear reactors pose greater security risks than conventional nuclear reactors do?

5. What makes nuclear power plants so costly to build?

6. How has deregulation of the U.S. electricity market led to the closing of some nuclear power plants?

7. What caused the nuclear accidents at Three Mile Island and Chernobyl?

8. Aside from health effects, in what ways have Ukrainians been affected by the Chernobyl disaster?

9. Describe the nature of the distribution of radioactive fallout from the Chernobyl accident.

10. What health effects have been observed in people exposed to radioactive fallout from Chernobyl?

11. Explain how exposure to radiation can lead to cancer.

12. What are some options for storing nuclear wastes, and which one of these options do most experts support?

Critical-Thinking Questions:

1. Imagine that you are trying to convince someone that nuclear power is better than fossil fuels as a way to generate electricity. What arguments will you make? Next, take the opposite position: Make a convincing argument that nuclear power is an unsuitable source of electricity.

2. Nuclear wastes must be safely stored for thousands of years as they slowly lose their radioactivity. Comment on the problems this longevity poses. Also, think of ways to ensure such long-term containment and transmit information to future generations. Keep in mind that other languages and written symbols may be in use in the future.

3. The nuclear accidents at both Three Mile Island and Chernobyl involved human error. Do you think computers should do more of the decision-making at nuclear power plants? In what other ways might human error be minimized?

Data Interpretation:

1. Plutonium-240 has a half-life of 6600 years. How long will it take for 75% of a plutonium-240 waste deposit to decay?

2. If 0.45 kg of uranium can release as much energy as 7300 metric tons of TNT, how much uranium (in kilograms) would be needed to release as much energy as 500 metric tons of TNT? How many pounds of uranium does your answer represent? How many ounces?

Chapter 12

Renewable Energy and Conservation

Study Outline:

I.   Direct Solar Energy
       A. Heating Buildings and Water
       B. Solar Thermal Electric Generation
       C. Photovoltaic Solar Cells
       D. Solar-Generated Hydrogen

II.   Indirect Solar Energy
       A. Biomass Energy
       B. Wind Energy
       C. Hydropower
       D. Ocean Waves
       E. Ocean Thermal Energy Conversion

III.   Other Renewable Energy Sources
       A. Tidal Energy
       B. Geothermal Energy

IV.   Energy Solutions: Conservation and Efficiency
       A. Energy Consumption Trends and Economics
       B. Energy-Efficient Technologies
       C. Electric Power Companies and Energy Efficiency
       D. Energy Conservation at the Individual Level

V.   Focus On: Cooking with Sunlight

VI.   Envirobrief: A Starring Role in Energy Efficiency

VII.   Envirobrief: Netting the Benefits of Home Energy Production

VIII. Envirobrief: Heating Goes Underground

IX.   Summary with Selected Key Terms

Key Terms: The following terms are listed in order of appearance in your textbook.

wind farms
renewable energy sources
solar energy
infrared radiation
passive solar heating
active solar heating
solar thermal electric generation
photovoltaic (PV) solar cells
fuel cell
biomass
wind energy
hydropower
biogas
biogas digesters
methanol
ethanol
schistosomiasis
Wild and Scenic Rivers Act
ocean temperature gradients
ocean thermal energy conversion (OTEC)
tidal energy
geothermal energy
hydrothermal reservoir
geothermal heat pumps (GHPs)
energy conservation
energy efficiency
energy intensity
National Appliance Energy Conservation Act (NAECA)
cogeneration
demand-side management
green power

Multiple Choice:

1. Which of the following involves direct solar energy?
   a. active solar heating
   b. passive solar heating
   c. solar thermal electric generation
   d. solar-generated hydrogen
   e. all of the above

2. Active solar heating is used primarily for
   a. cooking food
   b. heating homes
   c. heating hot water
   d. generating electricity
   e. heating commercial greenhouses

3. Sun-tracking mirrors guided by computers are part of the process of generating electricity through
   a. photovoltaic solar cells
   b. solar-generated hydrogen
   c. ocean thermal energy conversion
   d. energy intensity
   e. solar thermal electric generation

4. _____ is the energy choice in many remote areas, where applications include the pumping of water and grinding of grain.
   a. Photovoltaic solar cells
   b. Solar-generated hydrogen
   c. Solar thermal electric generation
   d. Hydropower
   e. Geothermal energy

5. Solar-generated hydrogen is a fuel obtained when electricity generated by photovoltaics or wind energy is used to split
   a. methane
   b. ammonia
   c. water
   d. ethanol
   e. biogas

6. When hydrogen is burned, it produces _____ and _____.
   a. heat, carbon dioxide
   b. heat, water
   c. sulfur oxides, water
   d. sulfur oxides, heat
   e. carbon monoxide, water

7. An advantage that solar-generated hydrogen has over photovoltaics is that it
   a. makes possible the storage of solar energy
   b. is less polluting
   c. is a cheaper source of electricity
   d. involves simpler technology
   e. is more available in rural areas

8. The type of biomass normally converted to biogas is
   a. peat
   b. animal wastes
   c. charcoal
   d. firewood
   e. fast-growing plants

9. Family-sized biogas digesters, used for heating and cooking, are common in
   a. Southeast Asia
   b. Japan and Great Britain
   c. Scandinavia
   d. India and China
   e. Portugal and Spain

10. The burning of biomass releases
    a. carbon dioxide
    b. sulfur dioxide
    c. methane
    d. hydrogen gas
    e. none of the above

11. Concerns about loss of habitat for native birds, including endangered species, led
    Sweden to cut back its projections of future _____ use.
    a. solar-energy
    b. wind-energy
    c. geothermal-energy
    d. biomass
    e. hydropower

12. _____ became the world's fastest growing source of energy during the 1990s.
    a. Solar energy
    b. Wind energy
    c. Geothermal energy
    d. Biomass
    e. Hydropower

13. Hydropower generates about _____% of the world's electricity.
    a. 10
    b. 20
    c. 30
    d. 40
    e. 50

14. Human-induced seismic activity is sometimes associated with this energy source:
    a. solar energy
    b. wind energy
    c. geothermal energy
    d. biomass
    e. hydropower

15. *Osprey*, the world's first commercial source of electricity produced by _____, sank in a storm less than 1 month after it opened.
    a. geothermal energy
    b. ocean thermal energy conversion
    c. ocean waves
    d. tidal energy
    e. photovoltaic solar cells

16. There is little commercial interest in _____ today.
    a. tidal energy
    b. photovoltaic solar cells
    c. ocean thermal energy conversion
    d. hydropower
    e. biomass

17. _____ heats most of the homes in Iceland.
    a. Passive solar heating
    b. Active solar heating
    c. Biomass
    d. Geothermal energy
    e. Ocean thermal energy conversion

18. There are few geographical locations suited to the use of
    a. passive solar energy
    b. tidal energy
    c. wind power
    d. hydroelectric power
    e. biomass

19. Making cars more fuel-efficient is an example of energy
    a. efficiency
    b. conservation
    c. intensity
    d. production
    e. all of the above

20. Japan has _____ than Europe and the United States.
    a. a less energy-efficient economy
    b. lower energy intensity
    c. higher per-capita domestic energy consumption
    d. lower energy costs
    e. all of the above

21. A major factor in the recent reduction of energy consumption in highly developed
    countries has been
    a. greater energy efficiency
    b. greater energy intensity
    c. declining population size
    d. global warming
    e. economic recession

22. A superinsulated home has most (or all) its windows on the
    a. roof
    b. north side
    c. east side
    d. south side
    e. west side

23. Typically, cogeneration involves the generation of _____ and the reuse of _____.
    a. heat, electricity
    b. steam, electricity
    c. light, heat
    d. light, steam
    e. electricity, steam

24. Air-conditioning costs can be reduced by
    a. using ceiling fans
    b. installing awnings and window shades
    c. planting trees
    d. using insulation
    e. all of the above

25. Demand-side management helps utility companies and their customers
    a. save money
    b. save energy
    c. avoid higher utility rates
    d. all of the above
    e. a and b only

26. Which of the following is *not* a source of green power?
    a. biomass
    b. solar power
    c. natural gas
    d. wind power
    e. hydroelectric power

27. Which of the following is *not* true of solar cookers?
    a. They are expensive to build.
    b. They can reach temperatures high enough for boiling.
    c. The technology is greatly underutilized.
    d. All of the above are untrue.
    e. Only a and b are untrue.

28. The Energy Star® labeling program offers consumers detailed energy-efficiency information about
    a. home heating systems
    b. automobiles
    c. appliances and computers
    d. their workplaces
    e. food preparation techniques

29. According to the EPA, _____ represent the most efficient heating system and release the least carbon dioxide.
    a. geothermal heat pumps
    b. gas furnaces
    c. wood stoves
    d. oil furnaces
    e. biogas digesters

Matching: Match the phrases on the left with the responses on the right. Some responses are used more than once.

_____ 1. typically associated with volcanism

_____ 2. convert sunlight directly to electricity

_____ 3. uses a flat solar panel or plate of black metal to absorb solar energy

_____ 4. provides fuel as a solid, liquid, or gas

_____ 5. based on temperature gradient of seawater; used to generate electricity

_____ 6. a result of the moon's and sun's gravitational pull

_____ 7. heat is stored in floors, walls, or containers of water

_____ 8. the most cost competitive of all forms of solar energy

_____ 9. can be used as a clean-burning auto fuel

_____ 10. a solar power tower is an example

_____ 11. can be incorporated into building materials as thin-films

_____ 12. will displace almost 2 million people in China

_____ 13. responsible for the spread of schistosomiasis in Egypt

_____ 14. oil or molten salt is heated to convert water to steam, which is used to generate electricity

_____ 15. one of the oldest fuels; based on chemical energy

a. passive solar heating

b. active solar heating

c. solar thermal electric generation

d. photovoltaic (PV) solar cells

e. solar-generated hydrogen

f. biomass energy

g. wind energy

h. hydropower

i. ocean thermal energy conversion

j. geothermal energy

k. tidal energy

Fill-In:

1. Arrays of wind turbines are called _____ (two words).

2. Invisible waves of heat energy are called _____ (two words).

3. _____ (three words) are thin layers of crystalline silicon treated with metals; they generate electricity when they absorb solar energy.

4. A _____ (two words), which combines oxygen and hydrogen, is an electrochemical cell similar to a battery.

5. Wood, crop wastes, and animal wastes are examples of _____, an organic source of energy.

6. Biomass can be converted to _____, often produced in family-sized _____ (two words). This fuel can be stored and transported in the manner of natural gas.

7. Biomass can be converted to liquid fuels, namely _____ and _____.

8. The Mesquite Lake Resource Recovery Project in southern California generates electricity by burning _____ (two words), a type of biomass.

9. What country is the world leader in wind power? _____

10. _____ is a tropical disease caused by a parasitic worm that has benefited from the building of the Aswan Dam.

11. The _____ (four words) Act prevents the building of hydroelectric dams on certain U.S. rivers.

12. Norway, Great Britain, and Japan are among the nations exploring the production of electricity from _____ (two words), an indirect form of solar energy.

13. What does OTEC stand for? _____

14. _____ (two words) is the natural heat produced in Earth's interior by radioactive decay and the friction of crustal plate movements.

15. Instead of hydrothermal reservoirs, _____ (three words) can be used as an expensive but more widely available source of geothermal energy.

16. Nova Scotia's Bay of Fundy has the largest _____ in the world.

17. _____ (two words) means reducing or stopping wasteful energy consumption.

18. _____ (two words) means using technology to accomplish a particular task with less energy.

19. _____ (two words) is a locality's total commercial energy consumption divided by its gross domestic product.

20. The _____ (four words) Act sets U.S. standards for the efficiency of washers, dryers, freezers, refrigerators, etc.

21. _____ is the production of two useful forms of energy from the same fuel.

22. Most utility companies offer _____ (two words) to help their customers reduce heating and air conditioning costs.

23. _____ (two words) allows homeowners with renewable energy systems to save money by contributing their excess energy to a utility's power grid and receiving in return energy–retail-price credit on their bills.

Short Answer:

1. Why does solar thermal electric generation often require a backup system, and what energy source does the backup system normally use?

2. What is a solar power tower, and how does it work?

3. What are some advantages and disadvantages of photovoltaic (PV) solar cells?

4. What are the disadvantages associated with biomass use?

5. What advantages and disadvantages are associated with wind energy?

6. How is hydropower linked to seismic activity?

7. How do geothermal heat pumps (GHPs) work, and how are they used?

8. Why do many energy experts favor energy conservation and efficiency as the best energy "sources"?

9. Why has U.S. average fuel efficiency of new cars not continued the improvement seen in the late 1970s and early 1980s?

10. Name some energy-saving changes that can be made to older homes.

11. What are some of the ways we, as individuals, can promote energy conservation?

Critical-Thinking Questions:

1. North America is currently experiencing major declines of freshwater invertebrate species, many of which are severely endangered. One reason given for these population nosedives is the building of hydroelectric dams, which greatly alter the ecology of the rivers they span. Do you think existing dams should be dismantled to restore rivers to their natural state and help rescue endangered species? Or, is the pressing need for energy adequate justification for the building of *new* dams? Is a compromise possible or desirable? Justify your response.

2. How can energy-efficient electric appliances slow the rate of global climate change?

3. What are some of the changes that could be made on your campus to reduce energy consumption?

4. Your text raises the challenge of promoting economic development without causing environmental damage, and it gives the example of Brazil choosing between building new power plants and investing in energy-efficient technologies. Which choice do you think is best? Why? Give another scenario in which a nation faces a choice that can either harm or benefit the environment, while providing for the nation's energy needs.

Data Interpretation:

1. The cost of manufacturing PV modules dropped from approximately $90 per watt in 1975 to about $4 per watt in 1998. What percentage drop in price does this change represent?

2. If every kWh of electricity generated by wind power instead of fossil fuels prevents 1.5 pounds of carbon dioxide from entering the atmosphere, how many *kg* of carbon dioxide are prevented from entering the atmosphere by generating 30,000 kWh of electricity through wind power? (See Appendix IV.)

3. By how much (in megawatts) did the generating capacity of global wind power increase between 1996 and 1997? (See Table A in *Environment*, 3/e.)

PART FIVE:

OUR PRECIOUS RESOURCES

Chapter 13

Water: A Fragile Resource

Study Outline:

I. The Importance of Water
    A. Properties of Water
    B. The Hydrologic Cycle and Our Supply of Fresh Water

II. How We Use Water
    A. Irrigation

III. Water Resource Problems
    A. Too Much Water

IV. Case in Point: The Floods of 1993
    A. Too Little Water

V. Water Problems in the United States
    A. Surface Water
    B. Mono Lake
    C. Groundwater

VI. Global Water Problems
    A. Drinking Water Problems
    B. Population Growth and Water Problems
    C. Sharing Water Resources among Countries

VII. Water Management
    A. Providing a Sustainable Water Supply

VIII. Case in Point: The Columbia River

IX. Case in Point: The Missouri River

X.   Water Conservation
         A. Reducing Agricultural Water Waste
         B. Reducing Water Waste in Industry
         C. Reducing Municipal Water Waste

XI.   Envirobrief: Saving Water by Xeriscaping

XII.   Envirobrief: Where Water Conservation Is Academic

XIII.   Envirobrief: U.S. Water Use Trends

XIV.   You Can Make a Difference: Conserving Water at Home

XV.   Summary with Selected Key Terms

Key Terms: The following terms are listed in order of appearance in your textbook.

reservoirs
hydrogen bond
vaporizes
sublimate
hydrologic cycle
surface water
wetlands
runoff
drainage basin
watershed
groundwater
aquifers
unconfined aquifers
water table
confined aquifer (or artesian aquifer)
arid
semiarid
flood plains
aquifer depletion
subsidence
saltwater intrusion
salinization
reclaimed water
Ogallala Aquifer
stable runoff
sustainable water use
desalinization (or desalination)
distillation

reverse osmosis
microirrigation (or drip irrigation, or trickle irrigation)
xeriscaped

Multiple Choice:

1. Approximately 97% of Earth's water is
   a. ice
   b. vapor
   c. fresh
   d. salty
   e. contained within the soil

2. Water is a _____ molecule that makes _____ bonds with other water molecules.
   a. polar, hydrogen
   b. polar, covalent
   c. polar, ionic
   d. nonpolar, hydrogen
   e. nonpolar, covalent

3. Water has a high
   a. melting/freezing point
   b. boiling point
   c. heat capacity
   d. all of the above
   e. none of the above

4. Water is densest when it is
   a. vapor
   b. boiling
   c. at room temperature
   d. just above its freezing temperature
   e. at or below its freezing temperature

5. A(n) _____ is an area of land drained by a single river.
   a. wetland
   b. drainage basin
   c. watershed
   d. unconfined aquifer
   e. confined aquifer

6. Aquifers are
   a. natural underground storage areas of fresh water
   b. types of wetlands
   c. drainage basins
   d. watersheds
   e. areas of saturated soil above the water table

7. _____ is the continent with the most irrigated agricultural land.
   a. Asia
   b. South America
   c. North America
   d. Africa
   e. Australia

8. Which of the following is *not* caused by aquifer depletion?
   a. lowering of the water table
   b. subsidence
   c. saltwater intrusion
   d. a potential threat to agriculture
   e. salinization

9. Most of the water that falls on the continental United States.
   a. makes its way to the oceans as runoff
   b. is used for irrigation
   c. is used for domestic and industrial purposes
   d. returns to the atmosphere through evaporation and transpiration
   e. recharges aquifers

10. Because of human activities, California's Mono Lake's water level _____ greatly and its salinity _____ dramatically.
    a. rose, increased
    b. rose, decreased
    c. dropped, increased
    d. dropped, decreased

11. The _____ River provides fresh water for parts of Mexico, several major U.S. cities (including Los Angeles, Denver, and Phoenix), agriculture, hydropower, and the Grand Canyon ecosystem.
    a. Mississippi
    b. Colorado
    c. North Platte
    d. Columbia
    e. Merced

12. The United States's Ogallala Aquifer is the largest _____ in the world. It is being depleted primarily by _____.
    a. surface water source, domestic and industrial uses
    b. surface water source, evaporation
    c. surface water source, agriculture
    d. groundwater deposit, domestic and industrial uses
    e. groundwater deposit, agriculture

13. The precipitation that falls in the Amazon River basin is not particularly useful to humans because
    a. it falls during only part of the year
    b. it causes extensive flood damage
    c. it falls on soils ill-suited for agriculture
    d. it is not enough to recharge depleted aquifers
    e. it promotes large mosquito populations, which limit human settlement

14. The World Health Organization estimates that 80% of human illness results from
    a. overcrowding
    b. saltwater intrusion
    c. water-borne parasites
    d. suppressed immune systems
    e. unclean and insufficient water supplies

15. _____ faces the most serious water shortages of any Western Hemisphere country.
    a. Canada
    b. Chile
    c. Mexico
    d. Haiti
    e. The United States

16. After years of abuse by industry and the dense population of its river basin, the _____ River is now almost as clean as drinking water and once again supports Atlantic salmon populations.
    a. Mississippi
    b. Rhine
    c. Ganges
    d. Euphrates
    e. Nile

17. People who live in the Aral Sea's watershed are susceptible to respiratory illnesses, anemia, kidney disease, and various cancers, all possibly related to the area's
    a. heavy industrialization
    b. toxic salt storms
    c. poor health care systems
    d. contaminated drinking water
    e. heritage of genetic diseases

18. In 1988, the Soviets buried hundreds of tons of _____ on an island in the Aral Sea.
    a. anthrax bacteria
    b. mercury
    c. radioactive wastes
    d. lead-based paint
    e. pesticides

19. Tensions between Israel, Jordan, the West Bank, and Gaza Strip are expected to increase as they compete for water from the
    a. Jordan River
    b. Dead Sea
    c. Tigris River
    d. Nile River
    e. Mediterranean Sea

20. Historically, _____ water use is most expensive and _____ water use is least expensive.
    a. industrial, agricultural
    b. agricultural, industrial
    c. industrial, domestic
    d. domestic, agricultural
    e. agricultural, domestic

21. In the spring of 1996, an experimental flood was used to help restore the _____ River ecosystem of _____ .
    a. Yellowstone, Yellowstone National Park
    b. Merced, Yosemite National Park
    c. Colorado, Grand Canyon National Park
    d. Greenbriar, Monongahela National Forest
    e. Salmon, Salmon National Forest

22. The many dams spanning the Columbia River have had a devastating effect on
    a. recreational opportunities
    b. commercial navigation
    c. irrigation in Washington
    d. migratory waterfowl
    e. salmon

23. The largest implementation ever of the Endangered Species Act involves _____ of the Northwest United States.
    a. beavers and river otters
    b. ground squirrels and pocket gophers
    c. salmon and steelhead trout
    d. freshwater clams and crustaceans
    e. aquatic vegetation

24. The longest river in the United States, the _____, contains North America's three
    largest reservoirs and is the subject of an intense water-rights conflict.
    a. Mississippi
    b. Tennessee
    c. Missouri
    d. Colorado
    e. Susquehanna

25. Much of southern California receives its water supply
    a. from desalinized saltwater
    b. via aqueducts from northern California and the Colorado River
    c. from a vast aquifer
    d. from individually owned wells
    e. from a large reservoir in Death Valley

26. Reverse osmosis is a method of
    a. salinization
    b. desalinization
    c. water conservation
    d. xeriscaping
    e. microirrigation

27. _____ is a huge industry in North Africa and the Middle East.
    a. Salinization
    b. Desalinization
    c. Water conservation
    d. Xeriscaping
    e. Microirrigation

28. In traditional flood-irrigation, more than half of the water applied to the soil normally
    a. evaporates into the atmosphere
    b. is absorbed by plants
    c. infiltrates the soil to the water table
    d. is lost as runoff
    e. is recycled for reuse

29. Recycling wastewater is an effective way to conserve _____ water use.
    a. agricultural
    b. municipal
    c. industrial
    d. a and c
    e. b and c

30. Two-thirds of water use in the average home occurs
    a. in the kitchen
    b. in the laundry room
    c. in the yard, watering plants
    d. in the bathroom
    e. outside, washing cars

Matching: Match the terms on the left with the responses on the right.

_g_ 1. wetland
_k_ 2. runoff
_n_ 3. groundwater
_d_ 4. water table
_i_ 5. flood plain
_l_ 6. Mississippi River
_m_ 7. Mono Lake
_e_ 8. Colorado River
_a_ 9. Ogallala Aquifer
_j_ 10. Rhine River
_f_ 11. Aral Sea
_h_ 12. Nile River
_b_ 13. Columbia River
_c_ 14. Missouri River

a. a U.S. resource; predicted to continue dropping
b. smolt migration downstream is hampered here
c. subject of a three-state suit against the Army Corps of Engineers
d. the upper limit of an unconfined aquifer
e. provides water for 20 million people; is saltier than the ocean in some places
f. has lost all 24 of its original fish species
g. an area of land covered with water for at least part of the year
h. ten nations share its basin
i. area bordering a river; receives overflow
j. suffered a severe chemical spill in 1986
k. replenishes surface waters
l. caused massive damage in 1993 floods
m. its water level is rising
n. flows slowly through permeable sediments or rocks

Fill-In:

1. An artificial lake that stores water for later use is called a _____.

2. Water _____ when it changes from a liquid to a gas; it _____ when it changes from solid to gas.

3. The _____ (two words) is the continuous circulation of water through the abiotic environment.

4. A _____ is the area of land drained by a river and all its tributaries.

5. A _____ (two words) is a groundwater storage area trapped between impermeable layers of rock.

6. Although _____ lands receive more precipitation than deserts (arid lands) do, they are still subject to significant droughts.

7. Runoff _____ (increases or decreases) significantly when an area is developed for human use.

8. About _____% of the world's human population lives in arid or semiarid lands.

9. The removal by humans of more groundwater than can be recharged by natural processes is called _____ (two words).

10. A phenomenon called _____ occurs when irrigation of arid and semiarid lands causes salt to build up in the soil.

11. _____ water is treated wastewater that is reused for irrigation, manufacturing, or some other purpose.

12. _____ runoff is the fraction of runoff from precipitation that is available throughout the year.

13. The United States _____ (four words) manages water projects in drought-prone areas of the world.

14. The beleaguered Aral Sea is in _____ (what country).

15. _____ management is more complicated and more difficult than surface water management.

16. _____ (three words) is the use of water in ways that do not threaten future supplies or interfere with the functioning of ecosystems or the hydrologic cycle.

17. The _____ (two words) Dam has greatly affected the Colorado River ecosystem in Grand Canyon National Park.

18. Young salmon are called _____.

19. _____ is one method of _____, or desalination; it involves vaporizing and then condensing water.

20. _____, also called drip or trickle irrigation, is an effective but expensive way to conserve water.

21. Yards that are _____ are planted with vegetation that requires very little water.

Short Answer:

1. What problems confront the San Francisco Bay and its delta?

2. Why is freshwater use increasing worldwide?

3. Why, in nature, is water never "pure"? Does this lack of purity mean water is unclean?

4. Is most groundwater considered a renewable or nonrenewable resource? Why?

5. Which two rivers have the largest watersheds?

6. What are some advantages of replacing large levees at river edges with smaller levees farther from the water?

7. What three practices exacerbated damage from the 1993 Mississippi River flood? Why?

8. How does water consumption (as a percent of renewable water) in the West and Southwest compare to water use in the rest of the United States?

9. How are the problems of California's Mono Lake and Kazakstan's Aral Sea similar?

10. Explain how it is possible for India to have high total runoff but low stable runoff.

11. In states that manage groundwater, what approaches are used?

12. Why were the decisions made to destroy dams on Maine's Kennebec River and Oregon's Sandy River?

13. What conflicts surround the use of water from the Missouri River?

14. What five industries consume almost 90% of industrial water in the United States?

15. What are some ways local governments can encourage water conservation?

16. Why did U.S. water use decrease by nearly 10% between 1980 and 1995?

Critical-Thinking Questions:

1. How is it that a channelized river can cause greater flood damage than one that is not channelized?

2. How can agriculture on marginal lands (those subject to frequent droughts) result in a vicious cycle of environmental degradation, poverty, and hunger?

3. How does microirrigation reduce the amount of salt left in irrigated soil?

4. Choose one of the many unresolved water conflicts presented in this chapter and devise an imaginative win-win solution that serves the needs of all people—and other species—involved.

Data Interpretation:

1. See the data given (Figure 13-2 in *Environment*, 3/e). Worldwide, how many times more water is used for irrigation than for domestic/municipal uses?

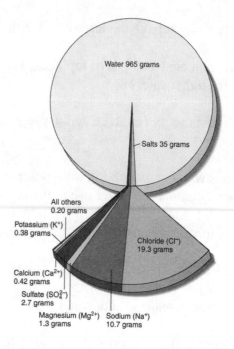

Water 965 grams

Salts 35 grams

All others 0.20 grams

Potassium (K$^+$) 0.38 grams

Chloride (Cl$^-$) 19.3 grams

Calcium (Ca^{2+}) 0.42 grams

Sulfate (SO$_4^{2-}$) 2.7 grams

Magnesium (Mg^{2+}) 1.3 grams

Sodium (Na$^+$) 10.7 grams

2. If 1 acre-foot of water provides enough water for 8 people/year, how many people are provided for by the 7.5 million acre-feet of water allotted annually to the lower Colorado by the 1922 Colorado River Compact?

3. See the figure (Figure 13-3 in *Environment*, 3/e) given. What percentage of Earth's *fresh* water is available for human use as groundwater?

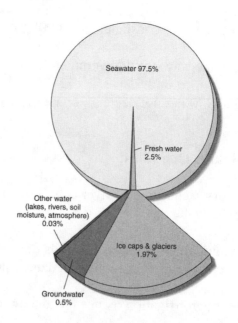

Seawater 97.5%

Fresh water 2.5%

Other water (lakes, rivers, soil moisture, atmosphere) 0.03%

Ice caps & glaciers 1.97%

Groundwater 0.5%

Chapter 14

Soils and Their Preservation

Study Outline:

I.   What Is Soil?
    A. How Soils Are Formed

II.   Soil Composition
    A. Components of Soil
    B. Soil Horizons
    C. Soil Organisms
    D. Nutrient Cycling

III.   Physical and Chemical Properties of Soil
    A. Soil Texture
    B. Soil Acidity

IV.   Major Soil Groups

V.   Soil Problems
    A. Soil Erosion

VI.   Case in Point: The American Dust Bowl
    A. Nutrient Mineral Depletion
    B. Soil Problems in the United States
    C. World Soil Problems

VII.   Soil Conservation and Regeneration
    A. Conservation Tillage
    B. Crop Rotation
    C. Contour Plowing, Strip Cropping, and Terracing
    D. Preserving Soil Fertility
    E. Soil Reclamation

VIII.   Soil Conservation Policies in the United States

Key Terms: The following terms are listed in order of appearance in your textbook.

desertification
weathering processes
topography
humus
soil water
soil air
leaching
illuviation
soil horizons
soil profile
O-horizon
A-horizon
E-horizon
B-horizon
C-horizon
castings
mycorrhizae
mycelium
nutrient cycling
sand
silt
clay
ions
loam
spodosols
alfisols
mollisols
aridisols
oxisols
sustainable soil use
soil erosion

agroforestry
conservation tillage
no-tillage
crop rotation
contour plowing
strip cropping
terracing
shelterbelts
Soil Conservation Act of 1935
Food Security Act (Farm Bill) of 1985
Federal Agriculture Improvement and Reform Act (Farm Bill) of 1996
compost
composted
mulch
municipal solid waste composting

Multiple Choice:

1. China's Great Green Wall was established to
   a. prevent Mongolian raids
   b. prevent dust storms
   c. reduce global warming
   d. provide wildlife habitat
   e. encourage ecotourism

2. Chemical and physical weathering processes generate _____ from _____.
   a. soil, rock
   b. rock, soil
   c. humus, organic matter
   d. organic matter, humus
   e. humus, clay

3. _____ and _____ play essential roles in weathering.
   a. Erosion, climate
   b. Leaching, organisms
   c. Organisms, climate
   d. Erosion, illuviation
   e. Leaching, illuviation

4. *Topography* is a region's
   a. weather
   b. history of volcanic activity
   c. organisms
   d. human history
   e. surface features

5. In general, older soils are _____ than younger soils in their _____ content.
   a. higher, essential-minerals
   b. lower, essential-minerals
   c. higher, water
   d. higher, humus
   e. lower, humus

6. Bacteria and fungi are the chief decomposers of _____, converting it to _____.
   a. sand, silt
   b. clay, humus
   c. organic "debris," humus
   d. humus, castings
   e. parent material, organic matter

7. Illuviation is the result of
   a. mycorrhizae
   b. decomposition of organic material
   c. weathering
   d. leaching
   e. all of the above

8. Generally, there is _____ carbon dioxide and _____ oxygen in soil air than in atmospheric air.
   a. less, more
   b. more, less
   c. less, less
   d. more, more

9. Light-colored subsoil—typically rich in iron, aluminum, and clay—is found in the _____ -horizon.
   a. O
   b. A
   c. E
   d. B
   e. C

10. The soil horizon most likely to be saturated with groundwater is the _____ -horizon.
    a. O
    b. A
    c. E
    d. B
    e. C

11. Earthworms _____ soil.
    a. eat
    b. aerate
    c. mix
    d. add organic matter to
    e. all of the above

12. Mycorrhizae help plants
    a. absorb minerals
    b. absorb water
    c. photosynthesize
    d. store energy
    e. avoid being eaten by herbivores

13. Leaching removes _____ from the soil.
    a. oxygen
    b. mineral nutrients
    c. microorganisms
    d. loam
    e. carbon dioxide

14. The smallest soil particles, which attract positively charged mineral ions, are _____ particles.
    a. clay
    b. sand
    c. loam
    d. silt
    e. parent-material

15. A(n) _____ provides the best aeration and drainage.
    a. sandy soil
    b. loam soil
    c. clay soil
    d. alfisol
    e. mollisol

16. Aluminum and manganese are examples of soil minerals that become more soluble (and potentially toxic to plants) in _____ soil, which has a _____ pH.
    a. basic, low
    b. basic, high
    c. acidic, low
    d. acidic, high
    e. neutral, neutral

17. A soil supporting the growth of _____ can be expected to be acidic.
    a. fungi
    b. maples
    c. grasses
    d. legumes
    e. conifers

18. _____ and _____ are both acidic, nutrient-poor soils.
    a. Oxisols, mollisols
    b. Alfisols, mollisols
    c. Aridisols, spodosols
    d. Spodosols, oxisols
    e. Alfisols, aridisols

19. Soil erosion
    a. limits plant growth
    b. diminishes soil fertility
    c. reduces agricultural productivity
    d. necessitates greater fertilizer use
    e. all of the above

20. _____ diversity plays a key role in promoting _____ diversity.
    a. Plant, fungal
    b. Fungal, plant
    c. Plant, soil
    d. Bacterial, plant
    e. Plant, bacterial

21. The Dust Bowl occurred during
    a. World War I
    b. the Roaring Twenties
    c. the Great Depression
    d. World War II
    e. the 1950s

22. On *all* farms, essential nutrients are lost through
    a. crop rotation
    b. soil erosion
    c. the harvesting of crops
    d. unsound plowing practices
    e. c and d

23. In a tropical rainforest, most of the ecosystem's nutrients are stored in the
    a. soil
    b. rocks
    c. vegetation
    d. groundwater
    e. surface water

24. One inch (2.5 centimeters) of lost topsoil typically takes _____ to replace.
    a. a few years
    b. tens of years
    c. hundreds of years
    d. thousands of years
    e. tens of thousands of years

25. Africa's Sahel is a region with
    a. overexploited, damaged soil
    b. deep, well-protected soil
    c. thin, yet highly productive, soil
    d. unproductive, water-logged clay soil
    e. fragile, rainforest soil

26. Increased _____ in the soil _____ the soil's water-holding capacity.
    a. organic matter, increases
    b. leaching, increases
    c. clay, decreases
    d. sand, increases
    e. mycorrhizae, decreases

27. The use of _____ lowers soil temperature, controls weeds, and reduces evaporation of water from the upper levels of the soil.
    a. compost
    b. mulch
    c. inorganic fertilizers
    d. crop rotation
    e. illuviation

28. _____ is likely to _____ in times of strong farm economy.
    a. Soil erosion, increase
    b. Soil erosion, decrease
    c. Leaching, increase
    d. Leaching, decrease
    e. Illuviation, increase

29. Municipal solid waste composting is limited in its success by concerns about _____ and _____ in the compost.
    a. bacteria, fungi
    b. fertilizers, pet wastes
    c. plastics, glass
    d. pesticide residues, heavy metals
    e. insect eggs, earthworms

Matching: Match the terms on the left with the responses on the right.

_____ 1. humus
_____ 2. leaching
_____ 3. illuviation
_____ 4. O-horizon
_____ 5. A-horizon
_____ 6. B-horizon
_____ 7. C-horizon
_____ 8. sand
_____ 9. silt
_____10. clay
_____11. loam
_____12. conservation tillage
_____13. crop rotation
_____14. contour plowing
_____15. shelterbelts

a. the topsoil
b. soil particles smaller than sand and larger than clay
c. reduces erosion by leaving crop residues in the soil
d. dark-colored organic material remaining after decomposition
e. the smallest soil particles; carry negative charges
f. rows of trees that protect soil from wind erosion
g. the removal of dissolved materials from the soil by water percolating downward
h. the subsoil; often a zone of illuviation
i. prevents erosion by avoiding furrows running up and down hills
j. soil particles from 0.05 to 2.0 mm in diameter
k. promotes soil fertility and helps prevent the accumulation of crop pests
l. the deposition of leached material in the soil's lower layers
m. contains plant litter; may be absent in desert soils
n. contains rocks and borders unweathered solid parent material (bedrock)
o. an excellent agricultural soil

Fill-In:

1. _____ is the degradation of productive lands into deserts.

2. More than _____% of human food comes from the land.

3. Although plants obtain most of the elements they need from the soil, they obtain _____ and _____ from the air.

4. Carbon dioxide forms _____ (two words) in water, which helps to weather rock.

5. _____, _____, and _____ are regions of the world with large areas of old, infertile soils.

6. A soil _____ is a section from the surface downward to the parent material; it reveals the soil _____.

7. _____ are little clumps of soil that have passed through an earthworm's gut.

8. The symbiotic relationship between fungi and plant roots is called _____. Fungal "threads," or _____, absorb minerals the plant needs; the plant provides the fungus with food.

9. _____ (two words) is the pathway of nutrient minerals or elements from the environment to organisms and back again.

10. The texture of the soil is determined by the relative amounts of _____, _____, and _____.

11. A soil that is too _____ loses its ability to hold positively charged mineral ions essential for plant growth.

12. Cool, wet areas of the world, dominated by coniferous forests, typically have soils called _____.

13. _____ are soils found in temperate deciduous forests.

14. _____ are fertile soils well suited to growing grains.

15. Deserts typically have soils called _____.

16. An Amazon-region soil, characterized by rapid decomposition of organic material and a highly leached B-horizon, is most likely a(n) _____.

17. _____ involves planting trees and crops together to improve degraded soils.

18. A type of conservation tillage called _____ leaves the soil largely undisturbed prior to and during the planting of seeds.

19. Alternating rows of corn and wheat, which follow the natural contours of the land, exemplify an erosion-control practice called _____ (two words).

20. Rice paddies in the Philippines and other mountainous areas are made possible through _____, which produces level areas.

21. Manure, bone meal, and compost are examples of _____ (organic or inorganic) fertilizers.

22. A natural soil and humus mixture that gardeners can produce themselves, at almost no expense, is _____.

23. The _____ (two words) Program pays farmers not to plant crops on erosion-prone soils.

Short Answer:

1. What does soil provide for plants?

2. What are the four components of all soils?

3. What is unusual about soils in the Everglades, and how are they threatened by agriculture?

4. How do ants benefit soils and plants?

5. What are the two types of organisms involved in the association known as *mycorrhizae*, and how do both benefit?

6. Why are there many thousands of soil types throughout the world?

7. How does soil erosion cause an increased need for fertilizer?

8. How does reducing soil erosion improve water quality?

9. How do plants prevent soil erosion?

10. Why can tropical rainforest soils be farmed successfully for only a few years?

11. What advantages do organic fertilizers, such as manure, have over inorganic fertilizers?

Critical-Thinking Questions:

1. Why do you think loams make such good agricultural soils?

2. Would the Dust Bowl have occurred if the same drought had taken place 100 years earlier? Justify your response.

Data Interpretation:

1. What is the total amount of soil, in billion metric tons, lost annually by India and China? This loss corresponds to how many billion metric tons per billion people per year?

2. In 1997, how many times more hectares of U.S. farmland were planted using conservation tillage than conventional plowing?

3. Annual loss of soil on CRP (Conservation Reserve Program) lands planted with grasses or trees has been reduced from 7.7 to 0.6 metric tons per hectare. What percent reduction in soil loss has occurred? Ten years at the "new" erosion rate corresponds to how much soil saved, per hectare, from eroding?

Chapter 15

Minerals: A Nonrenewable Resource

Study Outline:

I.   Uses of Minerals

II.   Mineral Distribution and Formation
      A. Formation of Mineral Deposits

III. How Minerals Are Found, Extracted, and Processed
      A. Discovering Mineral Deposits
      B. Extracting Minerals
      C. Processing Minerals

IV.   Environmental Implications
      A. Mining and the Environment
      B. Environmental Impacts of Refining Minerals

V.   Case in Point: Copper Basin, Tennessee
      A. Restoration of Mining Lands

VI.   Mineral Resources: An International Perspective
      A. U.S. and World Use
      B. Distribution Versus Consumption
      C. Will We Run Out of Important Minerals?

VII. Increasing Our Mineral Supplies
      A. Locating and Mining New Deposits
      B. Minerals in Antarctica
      C. Minerals from the Ocean
      D. Advanced Mining Technologies

Key Terms: The following terms are listed in order of appearance in your text.

General Mining Law
minerals
sulfides
oxides
rocks
ore
high-grade ores
low-grade ores
metals
nonmetallic minerals
magmatic concentration
hydrothermal processes
sedimentation
evaporation
surface mining
subsurface mining
overburden
open-pit surface mining
strip mining
spoil bank
shaft mine
slope mine
smelting
acid mine drainage
tailings
derelict lands
Surface Mining Control and Reclamation Act
phytoremediation
mineral reserves

mineral resources
total resources (or world reserve base)
life index of world reserves
Environmental Protection Protocol to the Antarctic Treaty (or Madrid Protocol)
manganese nodules
U.N. Convention on the Law of the Sea
reuse
recycling
sustainable manufacturing
dematerialization
industrial ecology
industrial ecosystems

Multiple Choice:

1. The $12.5- to $17.5-billion cost of cleaning up all 52 Superfund abandoned-mine sites in the West is paid by
   a. the companies that own the mines
   b. U.S. taxpayers
   c. users of the mined materials, priced to cover environmental costs
   d. local governments
   e. Worldwatch Institute

2. _____ is used for electrical and communications wiring.
   a. Steel
   b. Aluminum
   c. Tin
   d. Copper
   e. Iron

3. _____ is used to make plastics and fertilizer; it is also used to refine oil.
   a. Sand
   b. Manganese
   c. Tin
   d. Gypsum
   e. Sulfur

4. During the _____ Age, the first metal alloy, a mixture of tin and copper, was widely used.
   a. Stone
   b. Golden
   c. Iron
   d. Lead
   e. Bronze

5. Malleable, lustrous minerals, which are good conductors of heat and electricity, are
   a. high-grade ores
   b. low-grade ores
   c. metals
   d. nonmetallic minerals
   e. sulfides

6. _____ and _____ are minerals that are relatively abundant in the Earth's crust.
   a. Copper, chromium
   b. Aluminum, iron
   c. Nickel, gold
   d. Silver, mercury
   e. Bronze, steel

7. Cuba is recognized as the source of much of the world's
   a. gold
   b. silver
   c. nickel
   d. tungsten
   e. copper

8. Lead, zinc, gold, and silver commonly form deposits through
   a. evaporation
   b. magmatic concentration
   c. sedimentation
   d. hydrothermal processes
   e. condensation

9. Based on life indices given in Table 15-2 in *Environment*, 3/e, which mineral is
   expected to become scarce first?
   a. aluminum
   b. copper
   c. tin
   d. nickel
   e. zinc

10. Iron, copper, stone, and gravel are normally extracted by
    a. open-pit surface mining
    b. strip mining
    c. subsurface mining
    d. biomining
    e. smelting

11. Acid mine drainage can cause
    a. mutations leading to cancer
    b. severe air pollution
    c. severe water pollution
    d. soil erosion
    e. the risk of explosion

12. Sulfur, lead, and cadmium are air pollutants commonly emitted from
    a. tailings
    b. derelict lands
    c. spoil banks
    d. smelters
    e. slope mines

13. Copper Basin, Tennessee, was devastated by
    a. deforestation
    b. acid rain
    c. severe erosion
    d. mining activities
    e. all of the above

14. The construction of _____ is a slow but effective way to clean up reclaimed mining lands.
    a. chlorinated ponds
    b. wetlands
    c. spoil banks
    d. tailings
    e. culverts and channels

15. Nickel can be profitably removed from soil by a plant called
    a. jewelweed
    b. twist flower
    c. columbine
    d. lupine
    e. seedbox

16. The Bolivian government does not address the environmental nightmare of its mining district because
    a. of lack of awareness of the problem
    b. mining is Bolivia's major industry
    c. clean-up is too costly
    d. of citizens' apathy
    e. of lack of expertise in phytoremediation

17. The United States and other countries have stockpiles of chromium, platinum, titanium, and other metals because these metals
    a. can be exported for high profits
    b. are especially abundant
    c. are of special importance to industry and defense
    d. are in decreasing demand
    e. are believed to have special potential for future communications technologies

18. During the 21st century, the price of many minerals is expected to _____, while availability _____.
    a. increase, decreases
    b. increase, increases
    c. decrease, increases
    d. decrease, decreases
    e. stay the same, decreases

19. The Madrid Protocol places a 50-year moratorium on mineral exploration and development in Antarctica
    a. to avoid conflicts among nations competing for resources
    b. to protect the fragile polar environment
    c. because of a lack of technology for mining at extremely low temperatures
    d. to avoid the threat of plutonium extraction
    e. to avoid harming a thriving ecotourism industry

20. Concern about environmental impacts, along with prohibitive costs, inhibits exploitation of minerals
    a. in seawater
    b. in the Amazon Basin
    c. in derelict lands
    d. by strip mining
    e. in manganese nodules on the ocean floor

21. Sodium chloride (table salt), bromine, and magnesium can be profitably extracted from
    a. the ocean floor
    b. manganese nodules
    c. seawater
    d. the Amazon Basin
    e. Antarctic ice

22. Copper, gold, and phosphates are minerals that
    a. can be extracted without causing environmental destruction
    b. require strip mining to extract
    c. have become alarmingly scarce
    d. can be extracted efficiently by using microorganisms
    e. are declining in usefulness

23. Mining in many arid regions is limited by
    a. a shortage of skilled workers
    b. extreme heat
    c. the great amount of water needed for extracting and processing minerals
    d. lack of access to industrial markets
    e. all of the above

24. In recent years, copper wiring in telephone cables has been replaced by
    a. glass fibers
    b. plastic tubing
    c. aluminum wiring
    d. steel wiring
    e. nickel wiring

25. _____, which catalyzes many chemical reactions, is an example of a mineral that has
    no known substitute.
    a. Iron
    b. Tin
    c. Cobalt
    d. Platinum
    e. Aluminum

26. Recycling one _____ saves the energy equivalent of about 6 ounces of gasoline; it
    also reduces air pollution.
    a. aluminum can
    b. sheet of notebook paper
    c. glass bottle
    d. plastic bag
    e. paper bag

27. Cyanide heap leaching is an environmentally disastrous way of mining
    a. cobalt
    b. chromium
    c. silver
    d. titanium
    e. gold

28. In Denmark, surplus heat from a power plant is used to warm homes, greenhouses,
    and a fish farm. Such use of "wastes" in a beneficial way is one of the goals of an
    industrial
    a. complex
    b. park
    c. matrix
    d. ecosystem
    e. co-op

Matching: Match the terms on the left with the responses on the right.

_____ 1. phosphorus
_____ 2. oxides
_____ 3. ores
_____ 4. magmatic concentration
_____ 5. hydrothermal processes
_____ 6. sedimentation
_____ 7. evaporation
_____ 8. overburden
_____ 9. spoil bank
_____10. tailings
_____11. dematerialization
_____12. biomining

a. involve heated groundwater dissolving, recombining, and depositing minerals
b. layering that occurs as molten rock cools and solidifies
c. deposits of table salt, borax, and gypsum commonly form in this way
d. mineral compounds that contain oxygen
e. uses microorganisms to extract minerals
f. can occur when warm river water meets colder ocean water
g. the component of an ore that is discarded as waste
h. used in medicines, fertilizers, and detergents
i. rocks with a high enough concentration of a mineral for profitable mining
j. layers of soil and rock above a mineral deposit
k. the tendency of products to become smaller and lighter in weight as they evolve
l. a hill of loose rock dumped into a strip-mine trench

Fill-In:

1. The mining industry _____ (supports or opposes) reform of the General Mining Law of 1872.

2. _____ are elements or compounds found in Earth's crust.

3. Steel is a blend of _____ and other metals.

4. _____ are aggregates, or mixtures, of various minerals.

5. The South American country of _____ is particularly rich in copper deposits.

6. Deposits of iron, manganese, phosphorus, sulfur, and other minerals have been formed by the process of _____.

7. The discovery of _____ in Canada's Northwest Territories has led hundreds of companies to stake mining claims.

8. _____ involves melting ores at high temperatures to separate impurities from metal.

9. _____ (three words) refers to pollution of soil and water with toxic substances washed out of spoil banks.

10. A common impurity in many ores is _____, the same impurity that scrubbers remove from coal smoke.

11. Lands that have been degraded by mining are called _____ lands.

12. The main obstacle to restoring derelict lands is lack of _____.

13. More than 800 _____ systems have been constructed in Appalachia to neutralize acid from mining wastes.

14. A plant, such as twist flower, that absorbs and stores high quantities of a metal, is called a _____.

15. The U.S. must import virtually all of its _____, a mineral used to make paint pigments and certain types of steel.

16. Mineral _____ are mineral deposits that are currently profitable to extract.

17. Mineral _____ are deposits of low-grade ores that are not currently profitable to extract.

18. A mineral's _____ (two words) of world reserves is an estimate of how long it will take for known reserves to be exhausted.

19. Mineral-rich, potato-sized rocks on the ocean floor are called _____ (two words).

20. _____ of a glass bottle saves more energy than recycling does.

21. A _____ society places emphasis on repairing, instead of discarding, damaged products.

22. Minimization of industrial wastes, by converting them to useful products, is known as _____ (two words).

23. Dematerialization _____ (does or does not) guarantee reduced consumption and waste.

Short Answer:

1. What was the purpose of the General Mining Law of 1872, and how does the law encourage environmental destruction?

2. In what ways does gold mining harm the environment?

3. What are some of the techniques used to locate mineral deposits?

4. When compared with surface mining, what advantages and disadvantages does subsurface mining have?

5. How do mines cause environmental degradation?

6. What pollutants are released from smelters?

7. Is U.S. consumption of metals, as a fraction of world consumption, proportionate to population size? Explain.

8. What five countries are the world's top mineral producers?

9. Why is it difficult to predict future supplies of minerals?

10. What is the U.N. Convention on the Law of the Sea (UNCLOS), and what is its function?

11. What are some minerals now being recycled, and what are the advantages of recycling?

Critical-Thinking Questions:

1. When the Surface Mining Control and Reclamation Act of 1977 was passed, why do you think only *coal* mines were required to restore derelict lands? Do you think all mines should be included in the provisions of the law?

2. In some cases, plastics can be used in place of metals. Do you see this substitution as a permanent or temporary solution to the declining availability and increasing costs projected for many metals? Explain.

Data Interpretation:

1. The clean-up of Superfund mining sites in the West is expected to cost U.S. taxpayers approximately $15 billion. If there are 52 sites, what is the average cost per site?

2. If world population is 6 billion and the U.S. population makes up 4.6% of the total, what is the U.S. population?

3. If recycling 1 aluminum can saves the energy equivalent of 6 ounces of gasoline, how many cans must be recycled to save the energy equivalent of 1 gallon of gasoline?

Chapter 16

Preserving Earth's Biological Diversity

Study Outline:

I. How Many Species Are There?

II. Why We Need Organisms
    A. Ecosystem Services and Species Diversity
    B. Genetic Reserves
    C. Scientific Importance of Genetic Diversity
    D. Medicinal, Agricultural, and Industrial Importance of Organisms
    E. Aesthetic, Ethical, and Spiritual Value of Organisms

III. Endangered and Extinct Species
    A. Extinctions Today
    B. Endangered and Threatened Species
    C. Where Is Declining Biological Diversity the Greatest Problem?

IV. Human Causes of Species Endangerment
    A. Habitat Loss
    B. Exotic Species
    C. Overexploitation

V. Case in Point: Disappearing Frogs

VI. Conservation Biology
    A. Protecting Habitats
    B. Restoring Damaged or Destroyed Habitats
    C. Zoos, Aquaria, Botanical Gardens, and Seed Banks
    D. Conservation Organizations

VII. Conservation Policies and Laws
    A. Habitat Conservation Plans
    B. The U.S. Biological Resources Division
    C. International Conservation Policies and Laws

VIII. Wildlife Management
    A. Management of Migratory Animals

IX.   Case in Point: Arctic Snow Geese
         A. Management of Aquatic Organisms

X.   What Can We Do About Declining Biological Diversity?
         A. Increase Public Awareness
         B. Support Research in Conservation Biology
         C. Support the Establishment of an International System of Parks
         D. Control Pollution
         E. Provide Economic Incentives to Landowners and Other Local People

XI.   Envirobrief: Solving Crimes Involving Organisms

XII.   Envirobrief: Pollinators in Decline

XIII.   Focus On: Reintroducing Endangered Animal Species to Nature

XIV.   Meeting the Challenge: Wildlife Ranching as a Way to Preserve Biological
         Diversity in Africa

XV.   Summary with Selected Key Terms

Key Terms: The following terms are listed in order of appearance in your textbook.

species
biological diversity (or biodiversity)
species diversity
genetic diversity
ecosystem diversity
ecosystem services
extinction
background extinction
mass extinction
endangered species
range
threatened
fragmentation
adaptive radiation
biotic pollution
commercial harvest
bellwether species (or sentinel species)
conservation biology
in situ conservation
ex situ conservation
restoration ecology
artificial insemination

embryo transfer
seed banks
Endangered Species Act (ESA)
World Conservation Strategy
national conservation strategy
wildlife management
flyways
commercial extinction
bioaccumulate
wildlife ranching (or game farming)

Multiple Choice:

1. In 1999, President Clinton proposed that the _____ be removed from the threatened list.
   a. black-footed ferret
   b. Carolina parakeet
   c. American bison
   d. dusky seaside sparrow
   e. American bald eagle

2. Soil enrichment and the production of food and antibodies are major ways _____ benefit humans.
   a. bacteria and fungi
   b. nematodes
   c. insects
   d. mollusks
   e. plants and algae

3. Which of the following have been produced through genetic engineering?
   a. human insulin
   b. vaccines
   c. more productive farm animals
   d. agricultural products with longer shelf life
   e. all of the above

4. Quinoa and the winged bean are examples of
   a. endangered species
   b. threatened species
   c. underutilized, highly nutritious species
   d. genetically engineered species
   e. species used pharmaceutically

5. The main cause of extinction of Florida's dusky seaside sparrow was
   a. air pollution
   b. habitat loss
   c. predation
   d. sport hunting
   e. disease

6. An example of a species threatened, in part, by its need for a very large territory is the
   a. California condor
   b. blue whale
   c. Carolina parakeet
   d. black-footed ferret
   e. green sea turtle

7. The green sea turtle is an example of an endangered species
   a. with very specialized food requirements
   b. with a very low reproductive rate
   c. restricted to a small island habitat
   d. with a limited breeding area
   e. all of the above

8. The _____ is an animal threatened, in part, by its highly specialized feeding habits.
   a. blue whale
   b. green sea turtle
   c. giant panda
   d. dusky seaside sparrow
   e. golden toad

9. Migratory birds that winter in _____ are declining much faster than birds that winter in other habitats.
   a. tropical grasslands
   b. tropical dry woodlands
   c. deserts
   d. tropical rain forests
   e. subtropical wetlands

10. The few countries that hold most of Earth's biological diversity are _____ nations that _____ able to afford protective measures to protect their rich biological heritage.
   a. developing, are
   b. developing, are not
   c. highly developed, are
   d. highly developed, are not

11. The greatest threat to biological diversity is generally agreed to be
    a. pollution
    b. habitat destruction
    c. sport hunting
    d. the introduction of exotic species
    e. subsistence hunting

12. The term "biotic pollution" refers to
    a. causes of global warming
    b. pesticides
    c. exotic species
    d. ozone thinning
    e. all of the above

13. The island of Guam lost 9 of 12 species of forest birds because of introduced
    a. fire ants
    b. goats
    c. cats
    d. cane toads
    e. brown tree snakes

14. Farmers had exterminated the Carolina parakeet by 1920 because
    a. its feathers were sold for profit
    b. it competed with livestock for grain
    c. they (the farmers) were greatly bothered by its loud vocalizations
    d. it ate the fruit they grew
    e. of a black market in caged birds

15. Unregulated hunting (i.e., over-hunting) led to the extinction of the _____ and the near-extinction of the _____.
    a. passenger pigeon, American bison
    b. Carolina parakeet, prairie dog
    c. black-footed ferret, pocket gopher
    d. cheetah, African elephant
    e. snow leopard, American black bear

16. Even in certain pristine, protected environments, _____ (considered bellwether species) are declining, some to the point of extinction.
    a. amphibians
    b. reptiles
    c. marsupials
    d. shorebirds
    e. birds of prey

17. Chytrid is a _____ that kills _____.
    a. parasitic insect, birds
    b. bacterium, snakes and lizards
    c. fungus, frogs
    d. virus, humans
    e. roundworm, members of the dog family

18. Pesticides, ultraviolet light, and parasites have all been blamed for
    a. declining populations of migratory birds
    b. the near extinction of the black-footed ferret
    c. declining amphibian populations in mountainous regions of Costa Rica
    d. frog deformities
    e. puzzling crop failures in Europe

19. Captive breeding programs in zoos serve as an example of
    a. restoration ecology
    b. in situ conservation
    c. ex situ conservation
    d. genetic reserves
    e. ecosystem services

20. Which of the following is allowed in some wildlife refuges in the United States?
    a. drilling for oil
    b. mineral development
    c. hunting
    d. military exercises
    e. all of the above

21. The _____ is a beneficiary of a successful captive-breeding program.
    a. American black bear
    b. whooping crane
    c. prairie dog
    d. African elephant
    e. golden toad

22. _____ species have recovered enough to be removed from the U.S. endangered- and
    threatened-species lists; all listed species _____ equal funding.
    a. No, receive
    b. Few, receive
    c. Many, receive
    d. Few, do not receive
    e. Many, do not receive

23. A wildlife manager trying to increase populations of quail and ring-necked pheasant would want to increase the area of
    a. wetlands
    b. mowed grass
    c. weedy, early successional fields
    d. immature forests
    e. mature, old-growth forests

24. The commercial extinction of many species of _____ led to a worldwide hunting ban that went into effect in 1986.
    a. migratory waterfowl
    b. whales
    c. fishes
    d. migratory songbirds
    e. migratory grazing mammals

25. Conservationists argue that a minimum of _____% of Earth's land should be set aside as protected ecosystems.
    a. 5
    b. 10
    c. 15
    d. 20
    e. 25

26. The Suriname Biodiversity Prospecting Initiative pays local people of Suriname, South America, for
    a. drugs obtained from rainforest plants they identify
    b. trophy hunting rights
    c. caged birds, such as parrots, returned to the wild
    d. rainforest plants popular as houseplants in highly developed nations
    e. their work preventing poaching in parks and reserves

27. The second most common cause of imperilment (after habitat loss/degradation) of U.S. species of plants, mammals, and birds, is
    a. exotic species
    b. overexploitation
    c. pollution
    d. lack of genetic diversity
    e. overspecialization of habitat requirements

28. The dusky seaside sparrow and Hawaiian oo
    a. have become abundant because of successful captive-breeding programs
    b. have recovered enough to be delisted in the United States.
    c. have been moved from the United States endangered-species list to the threatened list
    d. are endangered
    e. are extinct

29. The Pacific gray whale, American alligator, and brown pelican
    a. have become abundant because of successful captive-breeding programs
    b. have recovered enough to be delisted in the United States.
    c. have been moved from the U.S. endangered-species list to the threatened list
    d. are endangered
    e. are extinct

30. Declines in bees and other pollinators are believed to be the result of
    a. diseases
    b. habitat alteration
    c. introduced pollinators
    d. pesticide use
    e. all of the above

Matching: Match the terms on the left with the responses on the right.

_____ 1. bald eagle	a. a beneficiary of embryo transfer
_____ 2. species diversity	b. any isolated habitat surrounded by unsuitable
_____ 3. ecosystem diversity	territory
_____ 4. major climate change	c. a worse threat to fishes, reptiles, and invertebrates
_____ 5. range	than to plants
_____ 6. island	d. a possible cause of mass extinction
_____ 7. fragmentation	e. an important part of in situ conservation
_____ 8. pollution	f. manages and conserves biological resources on
_____ 9. blue water hyacinth	U.S. federal lands
_____10. black-footed ferret	g. an example of biotic pollution in Florida
_____11. poaching	h. the number of species in natural communities
_____12. American black bear	i. strives to protect ecosystems and stabilize human
_____13. frogs	population
_____14. restoration ecology	j. harmed by the pesticide DDT
_____15. bongo	k. used to resolve conflicts between conservation
_____16. artificial insemination	and development interests
_____17. the ESA	l. creates "islands" and threatens many species
_____18. habitat conservation plans	m. illegal commercial hunting
_____19. Biological Resources	n. sentinel species harmed by UV light
Division (BRD)	o. the variety of interactions among organisms in
_____20. World Conservation	natural communities
Strategy	p. harmed by loss of prey
_____21. Arctic snow goose	q. a tool of ex situ conservation
_____22. feasibility study	r. too abundant on its summer range
	s. the area in which a particular species is found
	t. done prior to reintroduction of captive-bred species
	u. killed for its gallbladder
	v. one of the strongest pieces of environmental
	legislation in the United States

Fill-In:

1. Most species _____ (have or have not) been evaluated for their potential usefulness.

2. _____ (two words) are important environmental functions provided by organisms within ecosystems.

3. When the last member of a species dies, the species is _____.

4. The continuous, low-level extinction of species is called _____ (two words).

5. An episode of extinction affecting many species in a relatively short period of time is called a _____ (two words).

6. An _____ species is in imminent danger of extinction; a _____ species is not in imminent danger but is likely to become endangered in the foreseeable future.

7. Declining biological diversity in the United States is most serious in the states of _____, _____, and _____; of these three states, _____ has the most endangered species.

8. Although they occupy only 7% of Earth's surface, _____ (three words) support as many as 40% of Earth's species.

9. The evolution of many related species from one ancestral species is called _____ (two words).

10. Excluding Antarctica, agricultural lands occupy _____% of Earth's land area.

11. _____ are particularly vulnerable to the detrimental effects of exotic species.

12. _____ (two words) is the removal of live organisms from nature.

13. _____ (two words) is the scientific study and protection of biological diversity.

14. _____ (two words) conservation focuses on preserving biological diversity in nature, while _____ (two words) conservation focuses on preserving biological diversity in human-controlled environments.

15. _____ (two words) store millions of frozen plant embryos.

16. Countries that have signed the biological diversity treaty of the 1992 Earth Summit are required to develop a _____(three words).

17. Wildlife management generally focuses on maintaining a specific species' _____, while conservation biology strives to manage an entire _____.

18. Migratory waterfowl follow established routes, called _____, in their spring and fall migrations.

19. Many whales were hunted to _____ (two words), at which point it was unprofitable to continue hunting them.

20. _____ is the only nation that honors neither the global ban on commercial whaling nor the Southern Ocean Whale Sanctuary.

21. Only _____ out of every 10 reintroductions of captive-bred animals is successful.

22. _____ (two words), also known as _____ (two words), involves private ownership of wildlife, which may be viewed or hunted for the owner's profit.

Short Answer:

1. What are some of the ecosystem services provided by plants, animals, fungi, and microorganisms?

2. What problem is associated with new high-yield varieties of food crops? How can this problem be alleviated?

3. Give some examples of organisms that provide pharmaceutical compounds. Give examples of industrial products that come from plants and animals.
4. What characteristics make species susceptible to extinction?

5. Why are many species endemic to islands endangered?

6. What are some of the ways humans destroy and degrade habitats?

7. What are the major ways in which humans cause species endangerment?

8. Why does the pet trade take an especially high toll on wildlife populations?

9. What reasons are given for the global phenomenon of amphibian decline?

10. What factors limit the effectiveness of protected areas in preserving biological diversity?

11. What are the disadvantages of seed banks?

12. What are the chief criticisms of the Endangered Species Act (ESA)?

13. What is CITES? What is its main goal, and what limits its effectiveness?

14. Name some of the ways we can reverse the extinction trend.

15. Explain how wildlife ranching, compared to traditional agriculture, is less harmful to the environment.

Critical-Thinking Questions:

1. Make a list of all the organisms you have benefited from in the past 24 hours.

2. Have humans always viewed themselves as "masters" of the planet and felt justified in exploiting and killing other species? If not, how do you think such a way of thinking arose? What do you think is the origin of the contrasting viewpoint, which sees humans' role as that of caretakers and protectors of all life forms?

3. Many scientists believe a minimum of 10% of Earth's land should be protected in reserves and national parks. Do you agree or disagree? Also, how should protected lands be chosen? Back up your position on these questions.

Data Interpretation:

1. Use Table 16-1 in *Environment*, 3/e, to determine what percentage of U.S. species imperiled by human activities are plants. Repeat for invertebrates, then mammals.

2. If 1999 of every 2000 species that have ever lived are extinct, what percentage of Earth's all-time species total is alive today?

Chapter 17

Land Resources and Conservation

Study Outline:

I. Importance of Natural Areas

II. Current Land Use in the United States

III. Wilderness
    A. Do We Have Enough Wilderness?
    B. The National Wild and Scenic Rivers System

IV. National Parks
    A. Threats to Parks
    B. Natural Regulation

V. Wildlife Refuges

VI. Forests
    A. Forest Management
    B. Deforestation
    C. Tropical Forests and Deforestation
    D. Why Are Tropical Forests Disappearing?
    E. Why Are Tropical Dry Forests Disappearing?
    F. Boreal Forests and Deforestation
    G. Forests in the United States
    H. U.S. National Forests

VII. Case in Point: Tongass National Forest

VIII. Rangelands
    A. Rangeland Degradation and Desertification
    B. Rangelands in the United States
    C. Issues Involving U.S. Rangelands

IX.   Wetlands
    A. Coastlines

X.   Agricultural Lands

XI.   Suburban Sprawl and Urbanization

XII.   Land Use
    A. Public Planning of Land Use
    B. Management of Federal Lands

XIII. Conservation of Our Land Resources

XIV.   Focus On: A National Park in Africa

XV.   Meeting the Challenge: The Trout Creek Mountain Working Group

XVI.   Envirobrief: Ecologically Certified Wood

XVII. Envirobrief: Preserving Doe's Natural Labs

XVIII. Summary with Selected Key Terms

Key Terms: The following terms are listed in order of their appearance in your text.

California Desert Protection Act
nonurban (or rural) lands
ecosystem services
wilderness
Wilderness Act
Wild and Scenic Rivers Act
Land and Water Conservation Fund Act
natural regulation
transpiration
monocultures
ecologically sustainable forest management
wildlife corridors
selective cutting
shelterwood cutting
seed tree cutting
clearcutting (or even-age harvesting)
deforestation
tropical rain forests
tropical dry forests
slash-and-burn agriculture

scrub savanna
boreal forests
conservation easement
fibrous root system
carrying capacity
overgrazed
degradation
desertification
Taylor Grazing Act
Federal Land Policy and Management Act
Public Rangelands Improvement Act
wetlands
Emergency Wetlands Resources Act
Food Security Act
prime farmland
urbanization
wise-use movement
environmental movement
ecosystem management
riparian areas

Multiple Choice:

1. The California Desert Protection Act bans _____ in Death Valley and Joshua Tree
   National Parks and the Mojave National Preserve.
   a. mining
   b. cattle ranching
   c. recreational vehicle use
   d. motorcycle races
   e. all of the above

2. Approximately _____% of the land in the United States is privately owned.
   a. 25
   b. 40
   c. 55
   d. 70
   e. 85

3. Areas with _____ designation are given the greatest protection of all U.S. federal
   lands.
   a. wildlife refuge
   b. national park
   c. national forest
   d. wetland
   e. wilderness

4. Which of the following U.S. agencies holds the most land?
   a. Bureau of Land Management
   b. National Park Service
   c. Fish and Wildlife Service
   d. U.S. Forest Service
   e. Army Corps of Engineers

5. The most common landscape type found in national wilderness areas is
   a. wetlands
   b. arctic tundra
   c. prairies
   d. mountains
   e. coastlines

6. The _____ has more cultural/historical sites than wilderness sites.
   a. U.S. Forest Service
   b. National Park Service
   c. Bureau of Land Management
   d. National Wildlife Refuge System
   e. Nature Conservancy

7. The most heavily visited national park in the United States is _____ National Park.
   a. Grand Canyon
   b. Zion
   c. Yosemite
   d. Shenandoah
   e. Great Smoky Mountains

8. Since the late 1960s, _____ have greatly increased in Yellowstone National Park's northern range and have overgrazed the entire ecosystem.
   a. wild goats
   b. bighorn sheep
   c. mule deer
   d. elk
   e. ground squirrels

9. The National Wildlife Refuge System was established by President
   a. Wilson
   b. F. D. Roosevelt
   c. T. Roosevelt
   d. Eisenhower
   e. Truman

10. A forested area with a high rate of transpiration is likely to have more _____ than a deforested area.
    a. soil erosion
    b. precipitation
    c. strong winds
    d. droughts
    e. oxygen consumption

11. _____ is/are a more severe problem in a monoculture forest than in a natural forest.
    a. Soil erosion
    b. Pests and diseases
    c. Poaching
    d. Nest predation
    e. Air pollution

12. _____ is the method of harvesting trees criticized for destroying habitat and causing soil erosion.
    a. Seed tree cutting
    b. Selective cutting
    c. Clearcutting
    d. Shelterwood cutting
    e. Deforestation

13. _____ removes all mature, dead, and "undesirable" trees, over the course of three harvests and many years.
    a. Seed tree cutting
    b. Selective cutting
    c. Clearcutting
    d. Shelterwood cutting
    e. Deforestation

14. Which of the following is *not* caused by deforestation?
    a. extinctions
    b. droughts
    c. decreased soil fertility
    d. soil erosion
    e. reduced release of carbon dioxide

15. Brazil, Zaire, and Indonesia have almost half the world's
    a. tropical dry forests
    b. wetlands
    c. tropical rain forests
    d. boreal forests
    e. temperate forests

16. Most tropical deforestation is the result of
    a. commercial logging
    b. subsistence agriculture
    c. cattle ranching
    d. agriculture for crop export
    e. charcoal production

17. Tropical *dry* forests are destroyed primarily for
    a. subsistence agriculture
    b. fuel
    c. cattle ranching
    d. development
    e. commercial logging

18. _____ forests make up the world's largest biome.
    a. Tropical rain
    b. Tropical dry
    c. Temperate deciduous
    d. Temperate rain
    e. Boreal

19. National forests are used for
    a. logging
    b. mining
    c. grazing
    d. hunting and fishing
    e. all of the above

20. Overgrazing of rangelands contributes to soil erosion and
    a. desertification
    b. local climate change
    c. increased frequency of fire
    d. increased carrying capacity
    e. increased severity of storms

21. Taxpayers currently subsidize ranchers, who pay too little for
    a. grazing fees
    b. water
    c. grain
    d. insurance
    e. livestock

22. Contraception for female horses, burros, and deer is an effort to protect
   a. cattle
   b. rangelands
   c. the taiga
   d. freshwater wetlands
   e. humans from Lyme disease

23. In southeastern Oregon, populations of the threatened Lahontan cutthroat trout are
   increasing as a result of efforts by a coalition of ranchers, environmentalists, and
   government officials to
   a. stock streams with trout
   b. decrease runoff from highways
   c. improve the condition of riparian areas
   d. create fish ladders on area dams
   e. all of the above

24. Provisions of the Clean Water Act have done a better job of protecting _____
   wetlands than _____ wetlands.
   a. coastal, inland
   b. inland, coastal
   c. freshwater, saltwater
   d. freshwater, brackish
   e. nontidal, tidal

25. Approximately two-thirds of the world's human population lives within 150
   kilometers (93 miles) of a(n)
   a. freshwater marsh
   b. grassland
   c. boreal forest
   d. coastline
   e. estuary

26. U.S. farms are becoming _____ and _____ numerous.
   a. smaller, less
   b. smaller, more
   c. larger, less
   d. larger, more

27. Which farming region of the United States is considered most threatened by
   population growth and urban/suburban sprawl?
   a. the North Carolina Piedmont
   b. Oregon's Willamette Valley
   c. the Midwest Corn Belt
   d. California's Central Valley
   e. South Florida

28. Approximately _____ of the world's land area is used for agriculture.
    a. one-tenth
    b. one-fifth
    c. one-third
    d. one-half
    e. two-thirds

29. The wise-use movement promotes
    a. mining and other commercial exploitation in wilderness areas and national parks
    b. development of wetlands
    c. logging of all national forest lands
    d. a, b, and c
    e. none of the above

30. Certification of "green" wood is based on
    a. sustainability of timber resources
    b. socioeconomic benefits to local people
    c. preservation of wildlife habitat
    d. protection of watersheds
    e. all of the above

Matching: Match the terms on the left with the responses on the right.

_____ 1. ecosystem services   a. supports millions of subsistence farmers
_____ 2. wilderness   b. found in rangelands; prevent erosion
_____ 3. transpiration   c. protected zones connecting unlogged areas
_____ 4. wildlife corridors   d. expanding in some regions, such as Vermont
_____ 5. selective cutting   e. e.g., pest control and waste recycling
_____ 6. seed tree cutting   f. considered the United States's most endangered
       ecosystem
_____ 7. tropical dry forests   g. removes nearly all the trees
_____ 8. slash-and-burn   h. removes mature trees
       agriculture   i. the Corn Belt, for example
_____ 9. boreal forests   j. uninhabited by humans
_____10. temperate forests   k. found in Alaska's Tongass National Forest
_____11. old-growth forest   l. act as the oceans' nurseries
_____12. fibrous root systems   m. the chief source of industrial wood and wood fiber
_____13. coastal wetlands   n. found in India, Kenya, and Egypt, for example
_____14. prime farmland   o. provides moisture for clouds and precipitation
_____15. South Florida
       landscape

Fill-In:

1. The _____ (four words) Act, passed in 1968, protects rivers with unusual aesthetic, recreational, geologic, historical, or ecological value.

2. Many national park sites have been purchased with money provided by the _____ (five words) Act of 1965.

3. For more than 30 years, Yellowstone National Park has followed a controversial management policy called _____ (two words), which allows nature to take its course.

4. Photosynthesis removes _____ (two words) from the atmosphere.

5. _____ (two words) forest management seeks to conserve forests for both commercial and ecological benefit.

6. *Even-age harvesting* is equivalent in meaning to _____.

7. The gravest threat to the world's forests is _____.

8. When a forest is removed, local streams and rivers become _____ (more or less) likely to flood.

9. About _____% of the water roots absorb is evaporated into the atmosphere.

10. There is a strong statistical correlation between _____ (two words) and deforestation.

11. Commercial logging of tropical forests is practiced particularly in what region? _____

12. Deforestation for cattle ranching is practiced primarily in what region? _____

13. To see _____ forests, you need to travel to Alaska, Canada, Scandinavia, or northern Russia.

14. Over the next few decades, what part of the U.S. is expected to lose the most forests to agriculture and development? _____

15. Alaska's Tongass National Forest is a _____ (two words) forest that continues to be clearcut.

16. The _____ (two words) of a rangeland is the maximum number of grazing animals the ecosystem can sustain without deteriorating.

17. The United Nations Environment Programme estimates that 135 million people are in danger of displacement as a result of _____.

18. Wild _____ and _____ on public rangelands can be destructive to the ecosystem.

19. _____, lands that are transitional between aquatic and terrestrial ecosystems, provide many essential ecosystem services.

20. The _____ (WRP) provides financial incentives for landowners to restore and protect wetlands.

21. The _____ (three words), a "wise-use" organization, consists of developers and oil and gas company representatives, who want to drain and develop wetlands.

22. Korup National Park in _____ (country) has the richest _____ in Africa.

23. U.S. Department of Energy (DOE) weapons labs are surrounded by vast protected areas, totaling 2 million acres. These areas, which DOE has begun to sell, are called _____ (four words).

Short Answer:

1. What are some of the benefits derived from maintaining natural areas adjacent to urban and agricultural areas?

2. What four agencies manage most federally owned land?

3. What business interests typically oppose additions to the National Wilderness Preservation System?

4. Why are grizzly bear populations in U.S. national parks threatened?

5. What products and benefits do we obtain from trees and forests?

6. How do forests help to "even out" the flow of water during droughts and floods?

7. How can deforestation affect hydroelectric-power production?

8. What three agents (activities) are the major causes of deforestation? (List them in the order of their impact.)

9. Why have temperate forests expanded in parts of the eastern United States?

10. Why is road building such a controversial issue in national forests?

11. What practices are part of rangeland management?

12. What benefits do wetland provide?

13. What did Congress ask the National Research Council (NRC) to do in 1992? Why?

14. What information should be gathered prior to land-use decisions?

Critical-Thinking Questions:

1. Why do you think a single, large wildlife population is more likely than several small "island" populations to survive environmental stresses?

2. Why do you think hunting and fishing are allowed on some national wildlife refuges? Do you think they should be? Justify your opinion.

3. Explain how new housing developments have a "snowball effect," consuming more land than that purchased by the developer. How can this effect be minimized?

4. A total of 286.7 million visits were recorded in U.S. national parks in 1998. How does this figure compare to the U.S. population in 1998?

Data Interpretation:

1. If the *monthly* grazing fee was $1.35 per cow, and the federal government collected $13.5 million in 1999, how many cows grazed on federal lands?

2. It is estimated that 42 million hectares (104 million acres) remain of the lower 48 states' original 89.4 million hectares (221 million acres) of wetlands. What percent has been lost? If 104 million acres equal 42 million hectares, how many acres equal 1 hectare?

Chapter 18

Food Resources: A Challenge for Agriculture

Study Outline:

I. Human Nutritional Requirements

II. World Food Problems
    A. Producing Enough Food
    B. Poverty and Food: Making Food Affordable for the Poor
    C. Cultural and Economic Effects on Human Nutrition

III. Plants and Animals that Stand Between People and Starvation

IV. The Principle Types of Agriculture

V. The Effect of Domestication on Genetic Diversity
    A. The Global Decline in Domesticated Plant and Animal Varieties

VI. Increasing Crop Yields

VII. Case in Point: The Green Revolution
     A. Increasing Crop Yields in the Post–Green Revolution Era

VIII. Increasing Livestock Yields

IX. Food Processing and Food Additives
    A. Protection of the Consumer
    B. Are Food Additives Bad?

X. The Environmental Impacts of Agriculture
    A. Using More Land for Cultivation

XI. Solutions to Agricultural Problems
    A. A Move toward Sustainable Agriculture, a Substitute for High-Input Agriculture
    B. Making Subsistence Agriculture Sustainable and More Productive
    C. Genetic Engineering

Key Terms: The following terms are listed in order of appearance in your textbook.

Organic Food Production Act
carbohydrates
cell respiration
proteins
amino acids
essential amino acids
lipids
minerals
vitamins
undernourished
malnourished
marasmus
kwashiorkor
undernourished
overnourished
famine
world grain carryover stocks
food security
high-input agriculture
industrialized agriculture
yields
subsistence agriculture
shifting agriculture
nomadic herding
polyculture
domesticated
germplasm
green revolution
food additives
coloring agents

preservatives
antioxidants
nitrates
nitrites
N-nitroso compounds
degradation
sustainable agriculture (or alternative or low-input agriculture)
organic agriculture
integrated pest management (IPM)
genetic engineering
bycatch
ocean enclosure
open management
Magnuson Fishery Conservation Act
Magnuson-Stevens Fishery Conservation and Management Act
aquaculture
mariculture

Multiple Choice:

1. _____ are complex organic molecules needed in very small amounts by cells; _____
   are inorganic elements needed by cells.
   a. Proteins, lipids
   b. Carbohydrates, proteins
   c. Essential amino acids, vitamins
   d. Vitamins, minerals
   e. Minerals, sugars

2. A listless child with dry, brittle hair and a greatly swollen abdomen most likely
   suffers from
   a. protein deficiency
   b. kwashiorkor
   c. malnutrition
   d. all of the above
   e. none of the above

3. A diet high in saturated fats, sugar, and salt is often associated with
   a. undernutrition
   b. marasmus
   c. kwashiorkor
   d. protein deficiency
   e. overnutrition

4. Stockpiles of grain have _____ each year since _____.
    a. increased, 1967
    b. increased, 1987
    c. remained stable, 1977
    d. decreased, 1967
    e. decreased, 1987

5. Which of the following lists shows *increasing* efficiency of livestock converting grain to animal tissue?
    a. farmed fish, chicken, pork, beef
    b. beef, pork, chicken, farmed fish
    c. chicken, beef, farmed fish, pork
    d. pork, farmed fish, beef, chicken
    e. beef, chicken, farmed fish, pork

6. The greatest *number* of hungry people is in _____; the greatest *proportion* of hungry people is in _____.
    a. Asia, Africa
    b. Asia, Latin America
    c. Africa, Asia
    d. Africa, Latin America
    e. Latin America, Africa

7. _____, _____, and _____ provide about half the calories humans consume.
    a. Beef, potatoes, wheat
    b. Potatoes, wheat, soybeans
    c. Rice, wheat, corn
    d. Beef, wheat, corn
    e. Rice, poultry, wheat

8. In 1995, major fish kills occurred in North Carolina streams after a _____ spill.
    a. fertilizer
    b. pesticide
    c. hog manure
    d. grain
    e. germplasm

9. When plants and animals are domesticated, their genetic diversity _____ and their susceptibility to new types of disease organisms _____.
    a. increases, increases
    b. increases, decreases
    c. remains unchanged, decreases
    d. decreases, increases
    e. decreases, decreases

10. The Cotswold sheep is a
    a. genetically engineered domesticated animal
    b. wild animal with potential for human use
    a. source of meat raised in livestock factories
    b. domesticated animal with high genetic diversity
    c. a now-rare variety of sheep traditionally raised in England

11. High-yield varieties of crops require
    a. fertilizers
    b. pesticides
    c. mechanized equipment
    d. high-energy costs
    e. all of the above

12. This figure (Figure 18-7 in *Environment*, 3/e) shows that the greatest increases in U.S. grain yields occurred
    a. before 1970
    b. during the 1970s
    c. during the 1980s
    d. during the 1990s

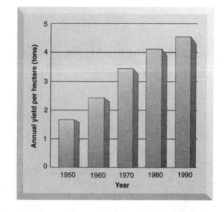

13. The EU (European Union) bans the import of _____ because of human health concerns.
    a. range-fed beef
    b. hormone-treated beef
    c. antibiotic-treated beef
    d. pesticide-treated corn
    e. pesticide-treated citrus fruit

14. _____ can combine with other chemicals to form carcinogenic compounds found in cured meats as well as smoked and salted fishes.
    a. Coloring agents
    b. Antioxidants
    c. Essential amino acids
    d. Nitrites
    e. BHT and BHA

15. _____ from animal wastes and fertilizers is/are a widespread groundwater
    contaminant in agricultural areas.
    a. Phosphates
    b. Silica
    c. Nitrites
    d. Sodium
    e. Carcinogens

16. Lester Brown of the Worldwatch Institute estimates that _____ tons of water are
    required to produce one ton of grain.
    a. 10
    b. 50
    c. 100
    d. 500
    e. 1000

17. Agriculture that relies on beneficial biological processes (such as natural predator-
    prey relationships) and environmentally friendly chemicals is called _____
    agriculture.
    a. sustainable
    b. low-input
    c. alternative
    d. high input
    e. a, b, and c

18. A 10-year study comparing agricultural methods showed _____ differences in corn
    yields; it also showed that only a _____ combination resulted in dramatically
    improved soil fertility.
    a. no significant, manure-corn-legume
    b. no significant, corn-legume
    c. significant, manure-corn-legume
    d. significant, corn-legume
    e. significant, manure-corn

19. An agricultural approach that uses a combination of pesticides, biological pest
    controls, crop rotation, careful monitoring for pests, and disease-resistant varieties is
    called
    a. subsistence agriculture
    b. high-input agriculture
    c. integrated pest management (IPM)
    d. shifting agriculture
    e. polyculture

20. The main problem with traditional shifting agriculture is that
    a. the soil is rapidly eroded
    b. pests are developing resistance to pesticides
    c. too many people are practicing it
    d. it requires fertilizers and expensive farming equipment
    e. burning trees causes air pollution

21. Genetic engineering differs from traditional breeding methods in that
    a. genes from one organism are transferred to another
    b. particular traits can be spread throughout a variety of plant or animal
    c. very rapid change in traits is possible
    d. a and b
    e. a and c

22. Incorporation into crop plants of a gene that prevents plants from absorbing aluminum (a potential plant toxin abundant in acidic soils) could provide great benefit for the people of _____, where 51% of the soil is acidic.
    a. North America
    b. Africa
    c. Northern Europe
    d. Latin America
    e. China

23. A very promising contribution of genetic engineering to livestock health involves
    a. protecting animals from cold weather
    b. creating new vaccines
    c. preventing birth defects
    d. increasing genetic uniformity
    e. making animals more docile

24. Collectively, humans obtain about 5% of their total dietary protein from
    a. poultry
    b. dairy products
    c. beef
    d. seafood
    e. eggs and beans

25. A _____ is used to catch bottom-feeding fishes, while a _____ is used to encircle large schools of fishes.
    a. longline, drift net
    b. trawl bag, purse-seine net
    c. drift net, trawl bag
    d. purse-seine net, longline
    e. trawl bag, longline

26. About _____% of all marine organisms caught are dumped, dead or dying, back into the sea.
    a. 10
    b. 15
    c. 20
    d. 25
    e. 30

27. _____ is the nation with the largest aquaculture harvest.
    a. China
    b. India
    c. Brazil
    d. Norway
    e. Canada

28. Which of the following may soon be prevented by edible vaccines (contained in GM plants)?
    a. diarrhea
    b. cholera
    c. tooth decay
    d. ulcers
    e. all of the above

29. Which of the following do vegans *not* eat?
    a. milk
    b. eggs
    c. seafood
    d. poultry
    e. all of the above

30. Which of the following food combinations provide(s) the proper complement of essential amino acids?
    a. corn and sunflower seeds
    b. rice and peas
    c. wheat and tofu
    d. all of the above
    e. b and c only

Matching I: For both matching sections below, match the phrases on the left with the responses on the right; responses may be used more than once.

_____ 1. metabolized readily in cell respiration
_____ 2. include fats, oils, and some hormones
_____ 3. components of proteins that must be provided by the diet
_____ 4. include sugars and starches
_____ 5. found in hair, nails, muscles
_____ 6. adenosine triphosphate
_____ 7. for humans, there are eight of these
_____ 8. large molecules made up of many amino acids; many are enzymes
_____ 9. by weight, they provide the most energy for cells
_____10. obtains energy from cell respiration

a. carbohydrates
b. lipids
c. essential amino acids
d. proteins
e. ATP

Matching II:

_____ 1. also called industrialized agriculture
_____ 2. practiced in arid areas
_____ 3. also called slash-and-burn agriculture
_____ 4. relies on large inputs of fossil-fuel energy
_____ 5. a form of subsistence agriculture; several crops are grown together
_____ 6. has led to replacement of family farms by agribusiness conglomerates
_____ 7. produces enough for oneself and one's family
_____ 8. makes heavy use of fertilizers and pesticides
_____ 9. one crop may fertilize others growing with it
_____10. a form of subsistence agriculture in which no chemical pesticides are used

a. nomadic herding
b. high-input agriculture
c. subsistence agriculture
d. organic agriculture
e. polyculture
f. shifting agriculture

Fill-In:

1. In 1990, the _____ (three words) Act established guidelines that define organic foods as crops grown in soil that has been free of chemical fertilizers and pesticides for at least _____ years.

2. Organic foods, grown without chemical fertilizers or pesticides, are currently _____ (more or less) expensive than foods grown by conventional methods.

3. People who do not receive as many calories as they need are said to be _____.

4. People who receive enough calories but not enough of some other nutritional requirement, such as protein or iron, are said to be _____.

5. The World Health Organization estimates that more than _____ billion people are malnourished. (Approximately what percentage of the total human population is malnourished?)

6. A disease called _____ is the result of a diet low in calories and protein.

7. Worldwide, more people die from _____ (famine or undernutrition/malnutrition).

8. According to the United Nations, a grain stockpile amount of _____ days is needed to safeguard world security. When was the most recent year that the world had such a quantity of grain reserves? _____

9. If a cow eats 2,500,000 kilocalories of grain before it is slaughtered, approximately how many kilocalories are stored in the cow? _____

10. The amount of a crop produced per unit of land is called the _____.

11. _____ is plant or animal material used in breeding.

12. The production of more food per acre, based on modern agricultural methods and the use of new, high-yielding varieties, is called the _____ (two words).

13. The International Food Policy Research Institute predicted in 1999 that world demand for rice, wheat, and corn will increase by _____% between 2000 and 2020.

14. Chemicals that enhance the taste, color, or texture of food, or improve nutritional value, prolong shelf life, or maintain food consistency are called _____ (two words); more specifically, chemicals that retard the growth of bacteria and fungi are called _____, chemicals that prevent oxidation are called _____, and chemicals that make food more visually appealing are called _____ (two words).

15. In the U.S., the monitoring of food additives is the responsibility of the _____ (four words).

16. What two tropical areas have the greatest potential for cultivation and cropland expansion? _____

17. The technique of taking a gene from one species and placing it into a cell of another species, where it is expressed, is called _____ (two words).

18. Unwanted fishes, dolphins, and other marine organisms, which are killed or injured by fishermen trying to catch *other* species, are collectively known as

    _____.

19. _____ (two words) is a policy that allows all of a nation's fishing boats unrestricted access to the waters under the nation's jurisdiction; _____ (two words), on the other hand, involves regulation of fishing in such waters.

20. The _____

    (five words) Act of 1996 requires—for U.S. marine fisheries—protection of habitat, reduction of overfishing, minimization of bycatch, and rebuilding of populations of overfished species.

21. _____ and _____ involve raising edible, aquatic organisms in enclosures.

22. Vegetarians must be careful to obtain a meal-by-meal balance of _____ (three words), which the body cannot store for future use.

Short Answer:

1. What human health problems are related to overnutrition?

2. During the past two decades, what countries have suffered the worst famines?

3. Why is it risky for humans to depend heavily on only a few species of genetically uniform crop plants?

4. What are some problems associated with the green revolution?

5. What risk is posed by routinely administering antibiotics to livestock?

6. What risks are posed by pesticide use?

7. Why has the amount of irrigated land in the United States decreased?

8. What techniques are increasing the productivity and longevity of New Guinea's agricultural forest plots?

9. Why was a GM (genetically modified) soybean *not* placed on the market?

10. What fishes are listed as "at risk" in the text? (See Table 18-5 in *Environment*, 3/e.) Why are ocean resources more susceptible to overuse and degradation than land resources are? And, why are fisheries experiencing increasing pressure?

11. By what means does the Magnuson-Stevens Fishery Conservation and Management Act strive to prevent overfishing in U.S. waters?

12. How does "shrimp farming" threaten the productivity and biodiversity of coastal waters?

Critical-Thinking Questions:

1. According to your text, 40% of the kilocalories (kcal) humans consume in highly developed countries comes from animals, while only 5% of the kcal consumed in developing countries comes from animals. How do you account for this difference?

2. If enough grain is produced to feed all the world's people, why is there so much hunger? What solutions can you think of?

3. Why do you think "coaxing more grain out of crops that were genetically improved during the green revolution has resulted in diminishing returns?" (See Figures 18-7 and 18-8 in *Environment*, 3/e.)

Data Interpretation:

1. By what percentage did worldwide per capita grain production decrease between 1984 (342 kg/person) and 1998 (314 kg/person)?

2. Per capita beef consumption is shown in Table 18-1 in *Environment*, 3/e. Assuming that these values have remained the same since they were calculated in 1996, use the data given, along with the *World Population Data Sheet* in the back of your text, to determine *total* beef consumption, in kilograms, in India, Italy, and the United States.

PART SIX:

ENVIRONMENTAL CONCERNS

Chapter 19

Air Pollution

Study Outline:

I.   The Atmosphere as a Resource

II.   Types and Sources of Air Pollution
        A. Major Classes of Air Pollutants
        B. Sources of Outdoor Air Pollution

III.   Effects of Air Pollution
        A. Air Pollution and Human Health

IV.   Urban Air Pollution
        A. How Weather and Topography Affect Air Pollution
        B. Urban Heat Islands and Dust Domes

V.   Controlling Air Pollutants

VI.   Air Pollution in the United States
        A. The Clean Air Act
        B. Other Ways to Improve Air Quality

VII.   Case in Point: Los Angeles

VIII.   Air Pollution in Developing Countries

IX.   Case in Point: Mexico City

X.   Long-Distance Transport of Air Pollution
        A. Movement of Air Pollution over the Ocean

XI.   Indoor Air Pollution
        A. Indoor Air Pollution and the Asthma Epidemic
        B. Radon
        C. Asbestos

Key Terms: The following terms are listed in order of appearance in your textbook.

air pollution
primary air pollutants
secondary air pollutants
particulate matter
solid particulate matter
nitrogen oxides
greenhouse gas
sulfur oxides
carbon oxides
hydrocarbons
ozone
hazardous air pollutants (or air toxics)
smog
industrial smog
photochemical smog
temperature inversion (or thermal inversion)
urban heat island
dust domes
Clean Air Act
global distillation effect
sick building syndrome
radon
Reading Prong
mesothelioma
sound

noise pollution
intensity
decibel (db)
decibel-A (dbA)
cochlea
hair cells

Multiple Choice:

1. _____ and _____ are constituents of the atmosphere that play a vital role in cell respiration and photosynthesis, respectively.
    a. Argon, nitrogen
    b. Carbon monoxide, carbon dioxide
    c. Oxygen, carbon dioxide
    d. Nitrogen, oxygen
    e. Water vapor, nitrogen

2. Ozone and sulfur trioxide are examples of
    a. primary air pollutants
    b. secondary air pollutants
    c. hydrocarbons
    d. air toxics
    e. solid particulate matter

3. Two hours of _____ use releases as much air pollution as driving a car 139,000 miles.
    a. farm tractor
    b. riding mower
    c. golf cart
    d. heat pump
    e. jet ski

4. Carbon monoxide is a breakdown product of _____, released by _____.
    a. isoprene, plants
    b. ozone, power plants
    c. nitrogen oxides, motor vehicles
    d. sulfur oxides, smelters
    e. carbon dioxide, animals

5. _____ and _____ reduce visibility, causing urban areas to receive less sunlight than rural areas.
    a. Particulate matter, ozone
    b. Hydrocarbons, nitrogen gas
    c. Carbon oxides, sulfur oxides
    d. Nitrogen oxides, hydrocarbons
    e. Sulfur oxides, air toxics

6. _____ contribute to both acid rain and global warming.
   a. Carbon oxides
   b. Sulfur oxides
   c. Nitrogen oxides
   d. all of the above
   e. only b and c

7. _____ contribute(s) to global warming.
   a. Carbon oxides
   b. Nitrogen oxides
   c. Hydrocarbons
   d. Ozone
   e. all of the above

8. _____ is/are harmful to plants.
   a. Nitrogen oxides
   b. Sulfur oxides
   c. Ozone
   d. all of the above
   e. only b and c

9. _____ is/are harmful to the respiratory tract of humans and other animals.
   a. Particulate matter
   b. Nitrogen oxides
   c. Sulfur oxides
   d. Ozone
   e. all of the above

10. Exposure to _____ can cause headaches, fatigue, sluggish reflexes—even death. It
    also aggravates the symptoms of congestive heart failure.
    a. carbon dioxide
    b. carbon monoxide
    c. sulfur dioxide
    d. sulfur trioxide
    e. ozone

11. Air pollution is a greater health threat to _____ (who breathe two times more air per
    pound of body weight) than to _____.
    a. adults, children
    b. children, adults
    c. adolescents, adults
    d. adults, adolescents
    e. the elderly, infants

12. Industrial smog is at it worst in _____, whereas photochemical smog is at its worst in
    _____.
    a. fall, spring
    b. spring, fall
    c. winter, summer
    d. summer, winter
    e. winter, spring

13. The main source of the ingredients of photochemical smog is
    a. motor vehicles
    b. bakeries
    c. dry cleaners
    d. home heating
    e. industrial emissions

14. Electric precipitators, fabric filters, and scrubbers reduce _____ pollution.
    a. ozone
    b. carbon-oxide
    c. particulate-matter
    d. sulfur-oxide
    e. methane

15. Catalytic converters can greatly reduce emissions of _____ and _____ by cars.
    a. sulfur oxides, carbon dioxide
    b. nitrogen oxides, carbon dioxide
    c. volatile hydrocarbons, carbon monoxide
    d. particulate matter, ozone
    e. sulfur oxides, methane

16. Air quality has not improved in some parts of the United States *primarily* because of
    a. an increase in the speed at which people drive
    b. an increase in the size of cars
    c. the failure of car manufacturers to produce cleaner-burning engines
    d. an increase in the number of cars
    e. failure of the EPA to enforce emissions regulations

17. Reducing the _____ content of gasoline reduces air pollution and raises the cost of
    gasoline.
    a. sulfur
    b. carbon
    c. nitrogen
    d. hydrogen
    e. oxygen

18. The dirtiest air in the United States is found in
    a. Baltimore
    b. New York
    c. Houston
    d. Chicago
    e. Los Angeles

19. The dirtiest air of any major city in the world is found in
    a. Calcutta, India
    b. London, England
    c. Tokyo, Japan
    d. Mexico City, Mexico
    e. Cairo, Egypt

20. The global distillation effect transports volatile chemical pollutants from _____ countries to _____ countries.
    a. warmer, colder
    b. colder, warmer
    c. more developed, less developed
    d. wetter, drier
    e. drier, wetter

21. Most arctic pollution is believed to come from industry and agriculture in
    a. China
    b. Scandinavia
    c. the U.S.
    d. Russia
    e. Canada

22. The number of people in the United States suffering from _____ has doubled since 1970.
    a. chronic bronchitis
    b. asthma
    c. pneumonia
    d. lung cancer
    e. emphysema

23. Exposure to radon increases the risk of _____.
    a. chronic bronchitis
    b. asthma
    c. pneumonia
    d. lung cancer
    e. emphysema

24. _____ greatly increases the risk of developing disease from asbestos exposure.
    a. The global distillation effect
    b. Ozone pollution
    c. A fatty diet
    d. Drinking alcoholic beverages
    e. Smoking

25. Excessive noise can cause
    a. permanent hearing loss
    b. increased blood pressure
    c. migraine headaches
    d. gastric ulcers
    e. all of the above

26. A panel convened by the National Institute of Environmental Health Sciences concluded that EMFs (electric and magnetic fields) should be considered a "possible human carcinogen" because exposure may cause
    a. lung cancer
    b. skin cancer
    c. leukemia
    d. heart disease
    e. gastric ulcers

27. Small spark-ignition (gasoline-powered) engines, used in lawnmowers and other lawn equipment, produce _____% of U.S. hydrocarbon emissions from mobile sources.
    a. 5
    b. 10
    c. 15
    d. 20
    e. 25

28. Hybrid cars use _____ to start the engine, _____for idling and low-speed driving, and both power sources for normal-speed driving.
    a. electricity, gasoline
    b. gasoline, electricity
    c. gasoline, alcohol
    d. hydrogen, solar power
    e. solar power, hydrogen

29. Passive smoking increases the risk of
    a. lung cancer
    b. respiratory infections
    c. allergies
    d. all of the above
    e. only a and b

30. _____ is declining in highly developed countries and increasing in many developing
    countries.
    a. The smoking habit
    b. Reliance on cars
    c. The incidence of asthma
    d. Noise pollution
    e. Exposure to EMFs

Matching: Match the phrases on the left with the responses on the right. Some responses
are used more than once.

_____ 1. a diverse group of organic compounds; one is methane   a. nitrogen oxides
_____ 2. can bring on asthma attacks and suppress the immune   b. sulfur oxides
          system   c. carbon oxides
_____ 3. probably the most serious indoor air pollutant   d. hydrocarbons
_____ 4. one type reduces the blood's ability to carry oxygen   e. ozone
_____ 5. has decreased the most since the Clean Air Act was   f. lead
          passed   g. radon
_____ 6. used in brake linings and electrical insulation   h. asbestos
_____ 7. damage metals and textiles
_____ 8. releases are reduced by no-tillage (a farming practice)
_____ 9. beneficial in the stratosphere; harmful in the troposphere
_____10. still added to gasoline in some developing nations
_____11. produced in larger quantities than any other air pollutant
_____12. can cause lung cancer and mesothelioma
_____13. released mostly by electric power plants
_____14. the most harmful component of photochemical smog
_____15. reached the United States from ore smelters in China
_____16. some are produced by plants, in response to heat

Fill-In:

1. The two most abundant gases in the atmosphere are _____ and
    _____.

2. _____ (two words) consists of gases, liquids, or solids in the
    atmosphere that harm organisms or materials.

3. _____ (three words) are harmful chemicals that
    enter the atmosphere directly.

4. _____ (three words) are harmful chemicals that
    form from other chemicals that have been released into the atmosphere.

5. The two main sources of primary air pollutants are _____ and
    _____.

6. The three industries that release the most toxic air pollutants are the
    _____, _____, and _____
    industries.

7. _____ (two words) consists of thousands of kinds of solids
    and liquids suspended in the atmosphere.

8. _____ (two words) pose long-term health risks to people
    exposed to high concentrations.

9. Urban air pollution that reduces visibility is called _____.

10. A _____ (two words) traps air pollution close to the ground.

11. Localized heat buildup, which occurs in cities, is called an _____
    (three words).

12. Urban heat islands contribute to the buildup of particulate matter, forming
    _____ (two words) over cities.

13. The _____ (two words) Act, first passed in _____,
    authorizes the EPA to set limits on specific air pollutants.

14. _____ was the first state to enforce emission standards on motor
    vehicles.

15. Worldwide, the leading cause of death for children is _____
    (two words).

16. The process that transports volatile chemicals from the tropics to higher latitudes is
    called the _____ (three words).

17. The presence of air pollution inside office buildings, where it adversely affects
    workers, is called _____ (three words).

18. The highest radon levels in the United States are found in homes on a geological
    formation, called the _____ (two words), found in parts of
    Pennsylvania, New Jersey, and New York.

19. Sound is called _____ when it becomes loud or disagreeable,
    particularly if it causes physiological or psychological harm.

20. Relative loudness (or intensity) of sound is expressed in units called
_____.

21. Prolonged exposure to noise impairs hearing by damaging _____
(two words) in the _____ of the ear.

22. Materials picked up by _____ (a type of insect) provide a source of
data about airborne pollutants.

23. Experimental cars with "reformers" use gasoline or methanol to produce
_____, which powers a fuel cell.

Short Answer:

1. How did Chattanooga, Tennessee, improve its air quality, once the worst in the
United States?

2. From a regulatory standpoint, what are the seven most important types of air
pollutants?

3. Why are microscopic types of particulate matter considered more dangerous than
larger types?

4. How do sulfur dioxide, nitrogen dioxide, and particulate matter affect the respiratory
tract?

5. What types of topography increase the risk of temperature inversions?

6. What six air pollutants has the EPA focused on since passage of the Clean Air Act in
1970?

7. The average sulfur content in gasoline is 330 ppm. What regions have imposed a new
limit of 40 ppm? Also, what states require emissions tests for diesel trucks and
buses?

8. Why is air quality deteriorating in many developing countries?

9. What are two major contributors to the particulate pollution that has plagued Mexico
City?

10. What are the most common contaminants of indoor air?

11. How is it possible for an electric car to release (indirectly) more emissions than a
gasoline-powered car?

12. Use Figure 19-3 in *Environment*, 3/e, to determine the largest sources of carbon monoxide, sulfur oxides, hydrocarbons, nitrogen oxides, and fine particulates.

Critical-Thinking Questions:

1. Despite California's strict auto emissions standards, reduced emissions from industrial and manufacturing sources, power plants that burn natural gas (the "clean" fossil fuel), and a subway system that reduces automobile use, Los Angeles still has the worst air pollution of any U.S. city. Why has Los Angeles found it so difficult to achieve clean air? What might be done to accelerate the progress being made? (See Figure 19-10 in *Environment*, 3/e, for sources of air pollution in L.A.)

2. How does air pollution contribute to water pollution?

3. City dwellers typically spend 90% to 95% of their time indoors. Estimate the percentage of time you have spent indoors during the past year. Look at your response to short-answer question #10. What can you do to reduce your exposure to indoor air pollutants?

Data Interpretation:

1. By how many million people did Mexico City grow between 1960 and 2000? By what percentage did the city grow during that four-decade time span?

2. Suppose 100 million cars, driven 10,000 miles per year, get an average of 25 mpg. How much gasoline would be saved per year if the average gas mileage was 80 mpg, the goal of the Partnership for a New Generation of Vehicles?

Chapter 20

Global Atmospheric Changes

Study Outline:

I.   Global Warming
    A. The Causes of Global Warming
    B. Developing Climate Models
    C. The Effects of Global Warming
    D. International Implications of Global Warming
    E. How We Can Deal with Global Warming

II.   Ozone Depletion in the Stratosphere
    A. The Causes of Ozone Depletion
    B. The Effects of Ozone Depletion

III.   Case in Point: Facilitating the Recovery of the Ozone Layer

IV.   Acid Deposition
    A. Measuring Acidity
    B. How Acid Deposition Develops
    C. Effects of Acid Deposition
    D. The Politics of Acid Deposition
    E. Facilitating the Recovery from Acid Deposition

V.   Links Among Global Warming, Ozone Depletion, and Acid Deposition
    A. Climate Warming May Hurt Ozone Recovery

VI.   Envirobrief: Developing Countries and Carbon Emissions

VII.   Envirobrief: Arctic Warming and the Native Cultures of the North

VIII.   Meeting the Challenge: Business Leadership in the Phase-Out of CFCs

IX.   Summary with Selected Key Terms

Key Terms: The following terms are listed in order of appearance in your textbook.

agroforestry
greenhouse effect
greenhouse gases
enhanced greenhouse effect
aerosols
aerosol effect
models
permafrost
Kyoto Protocol
carbon management
ultraviolet (UV) radiation
chlorofluorocarbons (CFCs)
Montreal Protocol
acid deposition
wet deposition
dry deposition
pH scale
forest decline

Multiple Choice:

1. Most human-made carbon dioxide comes from
   a. burning fossil fuels
   b. release of CFCs, which break down in the atmosphere
   c. burning forests
   d. decay of organic matter in landfills
   e. manufacturing industries

2. Methane is a greenhouse gas that comes from
   a. car exhaust
   b. smelters
   c. coal-fired power plants
   d. landfills and large mammals, especially cattle
   e. volcanic eruptions

3. Sulfur haze contributes to _____, which _____ the local climate.
   a. the greenhouse effect, cools
   b. the greenhouse effect, warms
   c. the aerosol effect, cools
   d. the aerosol effect, warms
   e. acid rain, warms

4. The eruption of Mt. Pinatubo in 1991 contributed to
   a. global cooling
   b. acid deposition
   c. the aerosol effect
   d. ozone depletion
   e. all of the above

5. Globally speaking, atmospheric carbon dioxide levels are _____ in summer than in winter because of _____.
   a. lower, increased photosynthesis
   b. higher, increased rates of decomposition
   c. lower, reduced heating needs
   d. higher, greater automobile use
   e. lower, increased phytoplankton populations

6. Based on data from 1996, _____ has the highest per-capita carbon-dioxide emissions of any nation.
   a. China
   b. the Russian Federation
   c. the United States
   d. Germany
   e. Japan

7. As a result of _____ evaporation, global warming is expected to cause more cloud cover. If these clouds are low-lying they will _____ the warming; if they are high cirrus clouds, they will _____ the warming.
   a. increased, increase, decrease
   b. increased, decrease, increase
   c. decreased, increase, decrease
   d. decreased, decrease, increase

8. Which of the following is *not* caused by global warming?
   a. altered patterns of precipitation
   b. reduced risk of flooding
   c. spread of certain tropical diseases
   d. shifting ranges of many species
   e. melting of permafrost

9. _____ is expected to _____ the frequency and intensity of storms, including hurricanes.
   a. Global warming, increase
   b. Global warming, decrease
   c. The aerosol effect, increase
   d. Ozone depletion, increase
   e. Ozone depletion, decrease

10. Declines in krill populations and increased _____ have diminished populations of Antarctica's Adélie penguins.
    a. drought
    b. snowfall
    c. leopard seal populations
    d. incidence of bird malaria
    e. sea level

11. 1998 was a devastating year for _____, of which about 10% died.
    a. mollusks
    b. insects
    c. tundra animal species
    d. corals
    e. wetland plants

12. The ocean conveyor belt, which _____ the North Atlantic, may be _____ by global warming.
    a. warms, weakened
    b. cools, weakened
    c. warms, strengthened
    d. cools, strengthened
    e. warms, reversed

13. Although developing nations are increasing their carbon emissions, highly developed nations currently release _____ times more carbon emissions per capita than developing nations.
    a. two
    b. three
    c. four
    d. five
    e. six

14. The Kyoto Protocol is an international agreement that provides timetables for reducing _____ emissions.
    a. sulfur dioxide
    b. nitric oxide
    c. greenhouse gas
    d. CFC
    e. methane

15. Taxes on greenhouse gases and carbon-dioxide-free technologies are strategies for dealing with carbon dioxide through
    a. adaptation
    b. mitigation
    c. sequestering carbon
    d. carbon management
    e. b and d

16. Fertilizing the ocean with iron increases the number of _____, which remove _____ from the atmosphere.
    a. zooplankton, carbon dioxide
    b. phytoplankton, carbon dioxide
    c. zooplankton, sulfur dioxide
    d. phytoplankton, sulfur dioxide
    e. phytoplankton, nitrogen oxides

17. The ozone layer is damaged primarily by
    a. carbon dioxide
    b. sulfur dioxide
    c. nitrogen oxides
    d. chlorofluorocarbons (CFCs)
    e. acid deposition

18. In 1995, Sherwood Rowland, Mario Molina, and Paul Crutzen received the first Nobel Prize given for environmental science research. Their work revealed
    a. the causes of global warming
    b. the causes of acid deposition
    c. the action of the ocean conveyor belt
    d. the causes of ozone depletion
    e. the effects of acid deposition on soil chemistry

19. Which of the following is *not* an effect of ozone depletion?
    a. mutations
    b. blindness
    c. weakened immunity
    d. loss of buffalo grass in the western United States
    e. declining Antarctic fish and phytoplankton populations

20. The Montreal Protocol is an international agreement to reduce emissions of
    a. sulfur dioxide
    b. nitric oxide
    c. greenhouse gases
    d. CFCs
    e. methane

21. _____ dioxide and _____ oxides form acids that _____ the pH of precipitation.
    a. Carbon, sulfur, lower
    b. Nitrogen, carbon, lower
    c. Sulfur, nitrogen, raise
    d. Nitrogen, carbon, raise
    e. Sulfur, nitrogen, lower

22. Fossil-fuel-burning power plants, smelters, and industrial boilers release which of the following?
    a. carbon dioxide
    b. CFCs
    c. nitrogen oxides
    d. sulfur dioxide
    e. a, c, and d

23. Germany's Black Forest, Nova Scotia's rivers, and lakes of the Adirondack region suffer particularly from
    a. acid deposition
    b. ozone depletion
    c. ozone pollution
    d. global warming
    e. regional droughts

24. As soils become more acidic, essential minerals, such as _____, become less available to plants; other potentially toxic minerals, such as _____, become more available.
    a. calcium, potassium
    b. manganese, aluminum
    c. calcium, aluminum
    d. hydrogen, nitrogen
    e. oxygen, hydrogen

25. Burning western low-sulfur coal reduces acid deposition but increases the release of _____ and _____ into the atmosphere.
    a. CFCs, arsenic
    b. nitrogen oxides, sulfur dioxide
    c. carbon dioxide, mercury
    d. ozone, lead
    e. carbon monoxide, hydrocarbons

26. A warming climate _____ organic material in lakes, which _____ the depth of UV penetration.
    a. increases, decreases
    b. decreases, increases
    c. increases, increases
    d. decreases, decreases

27. Greenhouse gases may be _____ the stratosphere, which _____ ozone destruction.
   a. cooling, increases
   b. cooling, decreases
   c. warming, increases
   d. warming, decreases
   e. thinning, increases

28. Which of the following changes, caused by global warming, has been observed by the Eskimo Inuit?
   a. warmer winters
   b. drying tundra
   c. thinning, retreating sea ice
   d. changes in wildlife numbers, distribution, and migration
   e. all of the above

29. McDonald's stopped using plastic-foam food containers made with CFCs when
   a. petitioned by schoolchildren
   b. sued by the EPA
   c. taken to court by outraged citizens
   d. it became cheaper to use paper containers
   e. threatened with negative publicity by the media

Matching: Match the terms on the left with the responses on the right. Some responses are used more than once.

____ 1. carbon dioxide	a. transport(s) heat around the globe
____ 2. methane	b. cause(s) skin cancer and cataracts
____ 3. nitrous oxide	c. contribute(s) to global warming
____ 4. UV radiation	d. mitigate(s) global warming
____ 5. tropospheric ozone	e. cool(s) the atmosphere
____ 6. chlorofluorocarbons (CFCs)	f. contribute(s) to acid deposition
____ 7. burning fossil fuels	g. block(s) UV radiation
____ 8. burning forests	
____ 9. aerosols	
____ 10. landfills and cattle	
____ 11. sulfur dioxide	
____ 12. volcanic eruptions	
____ 13. ocean conveyor belt	
____ 14. planting trees	
____ 15. stratospheric ozone	

Fill-In:

1. In the growing practice of _____, both forestry and agricultural techniques are used to improve degraded lands.

2. The natural trapping of heat by carbon dioxide and other greenhouse gases in the atmosphere is called the _____ (two words); additional warming caused by increased levels of greenhouse gases is called the _____ (three words).

3. Computers are used to run simulation _____, which calculate complex equations designed to represent the dynamics of climate.

4. A climate model is only as good as the _____ and _____ on which it is based.

5. According to the Worldwatch Institute, global _____ (two words) is starting to uncouple from carbon dioxide emissions.

6. Between 1900 and 2000, sea level rose about _____cm (_____in), mostly as a result of _____ (two words).

7. As the atmosphere warms, more water _____, which releases more _____ into the atmosphere, which generates more powerful _____.

8. Research revealed a(n) _____% decline in _____ populations in the California Current since 1951. This decline is believed to be the result of _____ (two words).

9. Certain weeds, insect pests, and disease-causing organisms are expected to expand their _____ and _____ in response to global warming.

10. Because of global climate changes, agricultural production may increase in _____ and _____.

11. Developing nations, which have produced _____% of carbon dioxide emissions since 1950, are projected to produce more carbon dioxide than highly developed countries by the year _____.

12. Energy consumption and carbon-dioxide emissions are largely a function of human _____ (two words).

13. _____ (two words) involves strategies for collecting carbon dioxide as it is released and sequestering it away from the atmosphere.

14. The construction of dikes and levees to protect coastal land is an example of _____ to global warming.

15. _____ radiation has wavelengths just shorter than those of visible light.

16. Ozone thinning was first observed over _____ (what land mass) in _____ (what year).

17. The _____ (two words) is a mass of cold air that circulates around the southern polar region, effectively isolating it from warmer air every fall (which is spring in the Southern Hemisphere); this is where chlorine and bromine break apart _____ molecules.

18. _____ (two words), the most dangerous type of skin cancer, is increasing faster than any other type of cancer.

19. A pH less than 7 is _____; a pH greater than 7 is _____.

20. In acidified soils, _____ ions replace many _____ (positively or negatively) charged mineral ions essential for plant growth.

21. Reduced vigor and health of trees, which leads to their eventual death, is called _____(two words); it is most pronounced at _____ (higher or lower) elevations.

22. Developing nations _____ (are/are not) making significant progress toward controlling greenhouse-gas emissions.

Short Answer:

1. What is the IPCC, and what is its role?

2. List present and future effects of global warming.

3. Name some countries particularly vulnerable to rising sea level because of large human populations in low-lying river deltas.

4. What ecosystems are considered to be at greatest risk of species loss from global warming?

5. Name some tropical diseases that are either already expanding their ranges or are feared to do so in the future, as global temperature rises.

6. What are the major criticisms of the Kyoto Protocol?

7. What mitigation strategies may lessen the severity of global warming?

8. Besides CFCs, what other compounds destroy ozone? What are the two chief types of chemicals used in place of CFCs, and how are they less than perfect?

9. In what way does acid rain harm birds?

10. What combination of stresses results in forest decline?

11. In what regions of the United States is rainfall less acidic than it was a few years ago, and why? And, why have engine improvements not been able to reduce nitrogen oxide emissions?

Critical-Thinking Questions:

1. How is it possible for global warming to cause both more frequent droughts and more frequent floods?

2. In response to pressure to reduce carbon emissions, India points out that it is the rich, industrialized nations, not developing nations, that have historically done most to cause global warming; therefore, why should India curb its emissions? Respond to this perspective; i.e., make a convincing argument that will encourage all developing nations to cooperate with efforts to reduce releases of carbon dioxide. Do you think highly developed nations owe developing nations financial and technical assistance?

Data Interpretation:

1. Carbon dioxide in the atmosphere rose from 280 ppm about 200 years ago to 367 ppm in 1998. What percentage of the 1998 carbon-dioxide level existed 200 years ago?

2. If burning a gallon of gasoline releases 5.5 pounds of carbon dioxide, how many pounds of carbon dioxide are released by a car that gets 28 mpg and is driven 130,000 miles in its lifetime? How many tons?

3. How many times more acidic is a pH of 3 than a pH of 5?

Chapter 21

Water and Soil Pollution

Study Outline:

I. Types of Water Pollution
    A. Sewage
    B. Disease-Causing Agents
    C. Sediment Pollution
    D. Inorganic Plant and Algal Nutrients
    E. Organic Compounds
    F. Inorganic Chemicals
    G. Radioactive Substances
    H. Thermal Pollution

II. Eutrophication: An Enrichment Problem in Standing-Water Ecosystems
    A. Controlling Artificial Eutrophication

III. Sources of Water Pollution
    A. Water Pollution from Agriculture
    B. Municipal Water Pollution
    C. Industrial Wastes in Water

IV. Groundwater Pollution

V. Improving Water Quality
    A. Purification of Drinking Water
    B. Municipal Sewage Treatment
    C. Individual Septic Systems

VI. Laws Controlling Water Pollution
    A. Safe Drinking Water Act
    B. Clean Water Act
    C. Laws That Protect Groundwater

VII. Case in Point: Water Pollution in the Great Lakes

Key Terms: The following terms are listed in order of appearance in your text.

water pollution
sewage
enrichment
cell respiration
biochemical oxygen demand (BOD) (or biological oxygen demand)
disease-causing agents
fecal coliform test
sediment pollution
inorganic plant and algal nutrients
hypoxia
organic compounds
inorganic
hypertension
radioactive substances
thermal pollution
oligotrophic
eutrophication
eutrophic
artificial eutrophication (or cultural eutrophication)
point source pollution

nonpoint source pollution (or polluted runoff)
combined sewer system
combined sewer overflow
reservoirs
primary treatment
primary sludge
secondary treatment
secondary sludge
tertiary treatment
Ocean Dumping Ban Act
Refuse Act
Safe Drinking Water Act
maximum contaminant level
Clean Water Act
national emission limitations
National Pollutant Discharge Elimination System
Resource, Conservation, and Recovery Act
Great Lakes Toxic Substance Control Agreement
soil pollution
salinization
soil remediation
dilution
vapor extraction
bioremediation
phytoremediation
nonylphenol
red tides

Multiple Choice:

1. In Arcata, California, sewage
   a. receives only primary treatment
   b. is treated by a series of wetlands
   c. goes directly into Humboldt Bay
   d. keeps wildlife out of the area's marshes
   e. is used to fertilize farms

2. Water polluted with _____ has a high BOD (biological oxygen demand).
   a. sediment
   b. sewage
   c. heavy metals
   d. radioactive substances
   e. disease-causing agents

3. Which of the following diseases is *not* transmitted by contaminated water?
   a. typhoid
   b. cholera
   c. tuberculosis
   d. bacterial dysentery
   e. salmonella poisoning

4. Presence of *Escherichia coli* in a water sample indicates contamination with
   a. sediment
   b. sewage
   c. heavy metals
   d. synthetic organic compounds
   e. motor oil

5. Sediments
   a. can clog the gills of aquatic animals
   b. reduce photosynthesis by aquatic producers
   c. can carry toxic pollutants into water
   d. can transport disease-causing agents into water
   e. all of the above

6. The Gulf of Mexico's "dead zone," the size of New Jersey, is the result of _____ causing _____.
   a. fertilizer and manure runoff, hypoxia
   b. sewage, a high BOD
   c. sediment, high turbidity
   d. mercury pollution, nervous system damage to fish
   e. pesticides, algal die-offs

7. Chloroform, vinyl chloride, and carbon tetrachloride are examples of synthetic organic pollutants that can cause
   a. nervous system damage
   b. blood disorders
   c. liver damage
   d. cancer
   e. kidney stones

8. Hypertension in men and hyperactivity in children are among the effects of high _____ levels in the body.
   a. magnesium
   b. mercury
   c. cadmium
   d. aluminum
   e. lead

9. _____ can get into the air in our homes when we shower or wash dishes and clothes with contaminated water.
   a. Lead
   b. Fecal coliform bacteria
   c. Mercury
   d. Radon
   e. Radioactivity

10. As water temperature _____, water holds less _____.
   a. rises, nitrogen
   b. rises, oxygen
   c. drops, nitrogen
   d. drops, carbon dioxide
   e. drops, oxygen

11. Abundance of _____ indicates a eutrophic body of water.
   a. catfish
   b. carp
   c. pike
   d. a and b
   e. a and c

12. Artificial eutrophication is caused by the addition of _____ to a body of water.
   a. pesticides
   b. disease-causing organisms
   c. nutrients
   d. toxins
   e. mollusks

13. According to the EPA's 1992 National Water Quality Inventory Report, _____ is/are the leading source of pollution of U.S. surface waters.
   a. industry
   b. lawn-care chemicals
   c. agriculture
   d. improperly treated sewage
   e. urban runoff

14. Salt, garbage, motor oil, and sediments are among pollutants typical of
   a. urban runoff
   b. industrial wastes
   c. agricultural runoff
   d. groundwater pollution
   e. sewage effluent

15. The electronics industry uses ion exchange and electrolytic recovery to remove _____ from wastewater.
    a. organic wastes
    b. radon
    c. plant nutrients
    d. heavy metals
    e. ozone

16. Pesticides, fertilizers, and organic compounds are the most common forms of
    a. industrial wastes
    b. point-source pollution
    c. urban-runoff pollution
    d. municipal water pollution
    e. groundwater pollution

17. Which of the following is *not* used to purify municipal water supplies?
    a. chlorine
    b. ozone
    c. ultraviolet light
    d. aluminum sulfate
    e. antibiotics

18. Much of Peru suffered a _____ epidemic in 1991, when chlorination of drinking water was discontinued.
    a. cholera
    b. typhus
    c. measles
    d. flu
    e. malaria

19. Fluoride is added to water to
    a. sterilize it
    b. make it clearer
    c. help prevent tooth decay
    d. make it taste better
    e. prevent it from corroding pipes

20. Many farmers would be willing to use sewage sludge to fertilize _____ but not _____.
    a. citrus fruits, sweet corn
    b. sweet corn, citrus fruits
    c. hay, vegetables
    d. vegetables, hay
    e. vegetables, feed-grain crops

21. What did New York City do with its sewage sludge prior to 1991?
    a. anaerobically digested it
    b. incinerated it
    c. dumped it in landfills
    d. dumped it in the ocean
    e. sold it to farmers

22. A major obstacle to the effective implementation of clean-water laws is
    a. citizens' apathy
    b. insufficient monitoring and enforcement
    c. loopholes in laws
    d. public protests
    e. lack of technological know-how

23. The establishment of drinking-water safety standards in the United States is primarily
    the responsibility of the
    a. Fish and Wildlife Service
    b. Department of Health and Human Services
    c. Environmental Protection Agency
    d. Department of the Interior
    e. National Security Agency

24. Which of the following is *not* an industrial city of the Great Lakes shoreline?
    a. Cincinnati
    b. Milwaukee
    c. Chicago
    d. Toronto
    e. Cleveland

25. The Great Lakes, which hold about one-fifth of the world's fresh surface water,
    formed when
    a. glaciers of the most recent ice age melted
    b. a vast wetland experienced eutrophication
    c. dams were built on the St. Lawrence River
    d. sea level rose and flooded the area
    e. earthquakes created vast depressions

26. An exotic species that has become a problem in the Great Lakes is the
    a. freshwater eel
    b. zebra mussel
    c. largemouth bass
    d. snapping turtle
    e. blue crab

27. _____, an element found in some western U.S. soils, can cause death and deformities in migratory birds when it gets into water.
    a. Iron
    b.) Selenium
    c. Magnesium
    d. Calcium
    e. Boron

28. A toxic heavy metal, found in certain fertilizers, that sometimes builds up in agricultural soils is
    a.) cadmium
    b. lead
    c. mercury
    d. aluminum
    e. cobalt

29. In a soil that is too salty, water moves out of _____ and into _____.
    a. roots, stems
    b. stems, roots
    c. the soil, groundwater
    d. the soil, roots
    e.) roots, the soil

30. Women's urine and a chemical called nonylphenol are suspected causes of
    a. fish kills
    b.) feminization of male fish
    c. groundwater pollution
    d. algal blooms
    e. soil salinization

31. Red tides, which can poison fish, water birds, marine mammals, and humans, are caused by
    a.) population explosions of certain marine algae
    b. die-offs of certain marine algae
    c. heavy-metal water pollution
    d. oil spills
    e. industrial dyes illegally dumped into the sea

Matching: Match the terms on the left with the responses on the right.

_f_ 1. enrichment
_g_ 2. hypoxia
_b_ 3. lead
_k_ 4. mercury
_n_ 5. radon
_j_ 6. chlorine
_a_ 7. primary treatment
_h_ 8. secondary treatment
_e_ 9. tertiary treatment
_d_ 10. salinization
_l_ 11. dilution
_c_ 12. vapor extraction
_m_ 13. bioremediation
_i_ 14. phytoremediation

a. removes suspended and floating matter from wastewater
b. can cause learning disabilities and attention deficit in children
c. involves use of air to remove volatile organic compounds from soil
d. can ruin soil after years of irrigation
e. removes nutrients, heavy metals, viruses, etc., from wastewater
f. the fertilization of a body of water
g. a condition of dissolved-oxygen depletion
h. uses microorganisms to remove organic matter from wastewater
i. makes use of plants to remove salts or heavy metals from soils
j. tentatively linked to rectal, pancreatic, and bladder cancers
k. released into the air when coal, plastics, batteries, and paints are burned; damages the nervous system
l. cleanses soil by rinsing it with large amounts of water
m. makes use of microorganisms to cleanse polluted soils and waters
n. a radioactive gas that increases the risk of lung cancer

Fill-In:

1. When organisms decompose sewage and other organic material, they do so through cell respiration, which requires (and consumes) _____.

2. The amount of oxygen needed by microorganisms to decompose wastes is called the _____ (three words).

3. The _____ (two words) test is used to check water samples for the presence of *Escherichia coli*, an indirect indication of sewage contamination.

4. Many organic compounds, such as pesticides and solvents, can be removed from wastewater through _____ treatment.

5. The single largest source of _____ (a heavy metal) pollution is coal-fired power plants.

6. A human-caused rise in temperature of a body of water is called _____ pollution.

7. A clear lake—with pike, sturgeon, and whitefish—is unenriched, or
    _____.

8. _____ (two words) occurs when human activities greatly
    accelerate the enrichment and filling in of a body of water.

9. The two nutrients generally responsible for eutrophication are _____
    and _____.

10. Agricultural runoff is an example of _____ source pollution;
    Europe's cyanide spill in February 2000 is an example of _____
    source pollution.

11. In a combined sewer system, a sewage treatment facility receives human and
    industrial wastes mixed with runoff from _____ (two words).

12. _____ (three words), which contains raw
    sewage, industrial wastes, and urban runoff, flows untreated into waterways.

13. The slimy mixture of particles and microorganisms that remains after secondary
    treatment of wastewater is called _____ (two words).

14. The anaerobic digestion of sewage sludge by microorganisms produces
    _____, which can be burned to heat the digesters (large circular
    tanks).

15. _____ (three words) are EPA restrictions on
    amounts of pollutants discharged into waterways by polluting industries.

16. The Resource, Conservation, and Recovery Act of 1976 deals with storage and
    disposal of _____ (two words).

17. Worldwide, at least _____ million people die of water-related illnesses each year.

18. Italy's Po River receives _____ from many cities, including Milan.

19. India's Ganges River is polluted with sewage, industrial wastes, and
    _____ (two words).

20. Kwale, Kenya, a site of cholera and diarrhea outbreaks in the past, now offers its
    residents clean, safe drinking water, thanks to the installation of a
    _____ (two words).

21. Many of the 4 million hand-pumped wells installed in Bangladesh are poisoning
    people with _____.

22. With a few exceptions, most soil pollutants originate as _____ chemicals, such as fertilizers and pesticides.

23. The clean-up of contaminated soil is called _____ (two words).

Short Answer:

1. How does sewage cause a high BOD (biological oxygen demand) in a body of water?

2. What are some common sources of sediment pollution?

3. Why did 370,000 people in the Milwaukee area develop diarrhea in 1993?

4. What are some sources of lead in our environment?

5. What are the possible effects of mercury (or methyl mercury) poisoning, and how are humans most often exposed to mercury?

6. What are some common water pollutants that come from agriculture?

7. What methods are used for sewage sludge disposal?

8. Why is continual monitoring by volunteers necessary to protect San Francisco Bay from polluters?

9. What federal laws protect water quality in the United States, and when were they first passed?

10. Name some signs of severe pollution of the Great Lakes (especially Lake Erie and Lake Ontario) in the 1960s.

11. What water-quality problems continue to plague the Great Lakes?

Critical-Thinking Questions:

1. More than 1 billion people lack access to safe drinking water. The health effects are devastating. What political, economic, or other effects might also result from this problem?

2. Federal environmental laws, such as the Clean Air Act, are under constant attack. For what reasons do some individuals and organizations resent or distrust such laws? How would a supporter of environmental laws respond to critics?

3. What are some of the safer substitutes for cleaning products? Why do you think so many people prefer the more expensive and more toxic cleaners?

Data Interpretation:

1. How much money was saved when Arcata, California, constructed a wetland wastewater treatment plant instead of a conventional one?

2. How many of the 1363 U.S. watersheds evaluated by the EPA by 1998 were seriously contaminated with toxic pollutants?

3. Based on the information given in this table (Table 21-3 in *Environment*, 3/e), how many times more dissolved oxygen can water hold at zero degrees (Celsius) than it can hold at forty degrees?

Table 21–3	DISSOLVED OXYGEN IN WATER AT VARIOUS TEMPERATURES

Temperature (°C)	Dissolved Oxygen Capacity* (g O_2 per L H_2O)
0	0.0141
10	0.0109
20	0.0092
25	0.0083
30	0.0077
35	0.0070
40	0.0065

*Data for water in contact with air at 760 mm mercury pressure.

Source: Joesten, M.D., and J.L. Wood. *World of Chemistry,*

Chapter 22

The Pesticide Dilemma

Study Outline:

I.   What Is a Pesticide?
   A. The "Perfect" Pesticide
   B. First-Generation and Second-Generation Pesticides
   C. The Major Groups of Insecticides
   D. The Major Kinds of Herbicides

II. Case in Point: The Use of Herbicides in the Vietnam War

III. Benefits of Pesticides
   A. Disease Control
   B. Crop Protection

IV. Problems Associated with Pesticide Use
   A. Development of Genetic Resistance
   B. Imbalances in the Ecosystem
   C. Persistence, Bioaccumulation, and Biological Magnification
   D. Mobility in the Environment
   E. Risks to Human Health

V.   Solutions to the Pesticide Dilemma
   A. Using Cultivation Methods to Control Pests
   B. Biological Controls
   C. Reproductive Controls
   D. Pheromones and Hormones
   E. Genetic Controls

VI. Case in Point: *Bt*, Its Potential and Problems
   A. Quarantine
   B. Integrated Pest Management

VII. Case in Point: Integrated Pest Management in Asia
   A. Irradiating Foods

VIII. Laws Controlling Pesticide Use
    A. Food, Drug, and Cosmetics Act
    B. Federal Insecticide, Fungicide, and Rodenticide Act
    C. Food Quality Protection Act

IX. The Manufacture and Use of Banned Pesticides
    A. The Importation of Food Tainted with Banned Pesticides
    B. The Global Ban of DDT

X. Changing Attitudes
    A. Pesticide Risk Assessment

XI. Envirobrief: A Lethal "Bug Juice"

XII. Envirobrief: Organic Cotton

XIII. Envirobrief: Pesticides in Schools

XIV. Focus On: The Bhopal Disaster

XV. Meeting the Challenge: Reducing Agricultural Pesticide Use by 50 percent in the United States

XVI. Summary with Selected Key Terms

Key Terms: The following terms are listed in order of appearance in your textbook.

pest
pesticides
insecticides
herbicides
fungicides
rodenticides
narrow-spectrum pesticide
broad-spectrum pesticide
botanicals
synthetic botanicals
first-generation pesticides
second-generation pesticides
dichloro-diphenyl-trichloroethane (DDT)
chlorinated hydrocarbon
organophosphates
carbamates
selective herbicides
nonselective herbicides

broad-leaf herbicides
grass herbicides
dioxins
pathogens
monoculture
genetic resistance
pesticide treadmill
resistance management
persistence
bioaccumulation (or bioconcentration)
biological magnification (or biological amplification)
biological controls
sterile male technique
pheromones
hormones
genetically modified (GM)
quarantine
integrated pest management (IPM)
Food, Drug, and Cosmetics Act (FDCA)
Pesticide Chemicals Amendment
Miller Amendment
Delaney Clause
Federal Insecticide, Fungicide, and Rodenticide Act (FIFRA)
Food Quality Protection Act
persistent organic pollutants (POPs)
calendar spraying
scout-and-spray

Multiple Choice:

1. Inorganic pesticides _____ widely used today; they are _____ toxic and tend to _____ in soil and water.
   a. are, mildly, break down
   b. are, highly, persist
   c. are not, highly, break down
   d. are not, highly, persist
   e. are not, mildly, persist

2. Plant-derived pesticides are called
   a. herbals
   b. herbicides
   c. botanicals
   d. carbamates
   e. organophosphates

3. The first of the second-generation pesticides was
   a. pyrethrin
   b. lindane
   c. DDT
   d. aldicarb
   e. Agent Orange

4. _____ are more poisonous than other classes of insecticides.
   a. Organophosphates
   b. Botanicals
   c. Carbamates
   d. Chlorinated hydrocarbons
   e. Narrow-spectrum pesticides

5. 2,4-D and 2,4,5-T are pesticides similar in structure to a natural growth hormone. They are used to kill
   a. insects
   b. rodents
   c. fungi
   d. nematodes
   e. plants

6. The danger of Agent Orange, a mixture of two herbicides used during the Vietnam War, came from minute concentrations of highly toxic
   a. hormones
   b. organophosphates
   c. arsenic
   d. carbamates
   e. dioxins

7. Pesticide use in many tropical countries is aimed at reducing _____ populations, which transmit _____ to humans.
   a. rodent, malaria
   b. mosquito, malaria
   c. mosquito, typhus
   d. rodent, typhus
   e. flea, AIDS

8. Maintaining "refuges" for pest species to avoid exposure to pesticides is a strategy of
   a. resistance management
   b. integrated pest management (IPM)
   c. biological control
   d. a and b
   e. b and c

9. Based on information given in Table 22-1 in *Environment*, 3/e, it appears _____ are most susceptible to pesticide poisoning.
   a. mammals
   b. fishes
   c. birds
   d. bees
   e. amphibians

10. DDT, endosulfan, and chlordane are examples of
    a. endocrine disrupters
    b. organophosphates
    c. carbamates
    d. narrow-spectrum pesticides
    e. synthetic botanicals

11. This graph (Figure 22-7 in *Environment*, 3/e) illustrates an example of
    a. creation of new pests
    b. bioaccumulation
    c. biological magnification
    d. genetic resistance
    e. pesticide persistence

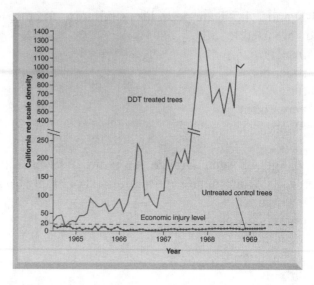

12. The novel chemical structure of synthetic pesticides, which prevents decomposers from degrading them, explains the _____ of many pesticides.
    a. severe toxicity
    b. effectiveness
    c. persistence
    d. broad-spectrum nature
    e. specificity

13. All top carnivores, including humans, are at particular risk from
    a. GM crops
    b. bioaccumulation
    c. biological magnification
    d. endocrine disrupters
    e. none of the above

14. In 1994, the Environmental Working Group found that 14.1 million U.S. residents drink water that contains small amounts of five widely used
    a. endocrine disrupters
    b. insecticides
    c. herbicides
    d. rodenticides
    e. fungicides

15. Studies of farm worker and pesticide-factory workers show a connection between _____ and long-term pesticide exposure.
    a. heart disease
    b. anemia
    c. diabetes
    d. asthma
    e. cancer

16. A study of Yaqui Indian preschoolers showed what effect(s) of pesticide exposure?
    a. suppressed immune systems
    b. deficiency of physical skills
    c. deficiency of mental skills
    d. a and b
    e. b and c

17. Cultivation methods that control pests include
    a. interplanting
    b. strip cutting
    c. crop rotation
    d. use of insect vacuums
    e. all of the above

18. An expensive but effective technique used to control medflies (Mediterranean fruit flies) is
    a. crop rotation
    b. use of insect vacuums
    c. genetic engineering
    d. the sterile male technique
    e. biological control

19. _____ are chemicals released by an organism that have an effect on another individual of the same species; _____, however, regulate the physiology of the organism that produces them.
    a. Pheromones, hormones
    b. Hormones, pheromones
    c. Organophosphates, hormones
    d. Organophosphates, pheromones
    e. Enzymes, hormones

20. MIMIC is a synthetic version of an insect hormone. It kills
    a. adult butterflies and moths
    b. larvae (caterpillars) of butterflies and moths
    c. beetles, including Japanese beetles
    d. ladybugs
    e. aphids

21. A gene from the bacterium *Bacillus thuringiensis* has been added to _____; it may
    (unintentionally) make the pollen toxic to _____.
    a. corn, monarch caterpillars
    b. citrus fruit trees, medflies
    c. tomatoes, songbirds
    d. melons, rodents
    e. wheat, honeybees

22. Farmers using IPM typically use pesticides
    a. on a regular basis
    b. when a pest population reaches an economic threshold
    c. just before a crop is harvested
    d. only when all other methods have failed
    e. never

23. _____ account(s) for almost 50% of all the insecticides used in U.S. agriculture.
    a. Wheat
    b. Corn
    c. Potatoes
    d. Cotton
    e. Soybeans

24. A study in Asia showed that rice fields treated with pesticides had yields _____
    untreated fields.
    a. the same as or smaller than
    b. the same as or larger than
    c. many times larger than
    d. many times smaller than
    e. identical to

25. Irradiating food
    a. kills microorganisms
    b. can reduce pesticide use
    c. extends its shelf life
    d. lessens the need for food additives
    e. all of the above

26. Approved by Congress in 1958, the _____ stated that no substance capable of causing cancer in test animals or humans would be permitted in processed food.
    a. Miller Amendment
    b. Pesticide Chemicals Amendment
    c. Delaney Clause
    d. Federal Insecticide, Fungicide, and Rodenticide Act
    e. Food Quality Protection Act

27. Currently, DDT is used in 23 countries to
    a. suppress insects that damage crops
    b. suppress weeds that compete with crops
    c. suppress bacterial pathogens that damage crops
    d. control malaria by killing mosquitoes
    e. control malaria by killing fleas

28. Using a blender to mix water and hornworms infected with a virus creates a safe and inexpensive way to protect _____ from hornworms.
    a. corn
    b. bananas
    c. citrus fruits
    d. cassava
    e. soybeans

29. In 1984, A toxic cloud of methyl isocyanate (MIC) gas and hydrogen cyanide, released from a Union Carbide pesticide plant, killed thousands of people in
    a. Madrid, Spain
    b. Bhopal, India
    c. Lawrence, Kansas
    d. Mexico City, Mexico
    e. Oslo, Norway

30. By 1992, _____ had reduced pesticide use by 50%. This nation is now working to reduce pesticide use by another 50%.
    a. Mexico
    b. the United States
    c. Poland
    d. China
    e. Sweden

Matching: Match the terms on the left with the responses on the right.

_____ 1. synthetic botanicals
_____ 2. chlorinated hydrocarbons
_____ 3. organophosphates
_____ 4. carbamates
_____ 5. monoculture
_____ 6. DDT
_____ 7. pesticides
_____ 8. 2,4-D
_____ 9. endocrine disrupters
_____ 10. biological controls
_____ 11. vedalia beetle
_____ 12. *Bt* corn
_____ 13. irradiating foods
_____ 14. persistent organic
       pollutants (POPs)

a. banned in the United States in 1972
b. malathion belongs to this category of insecticide
c. include predators, parasites, and disease-causing organisms
d. pesticides derived from plant compounds; pryethroids, for example
e. may cause reproductive problems in animals, including humans
f. the most toxic chemicals on Earth
g. most are broad-spectrum insecticides; few are still in use
h. poison over 3 million people per year
i. herbicide linked to a type of lymphoma
j. genetically modified to produce a toxin
k. can prevent food poisoning
l. a category of insecticide safer than many for mammals
m. promotes growth of pest populations
n. used successfully to protect citrus trees from the cottony-cushion scale

Fill-In:

1. A _____ pesticide kills only one organism, the target pest; most pesticides, however, are _____, meaning they kill other organisms along with the pest.

2. _____ herbicides kill all plants; _____ herbicides kill only certain types of plants.

3. Because of the spraying of Agent Orange during the Vietnam War, the _____ content of breast milk of Vietnamese women has been measured at 1800 parts per trillion (as compared with the U.S. average of 4 parts per trillion).

4. Microorganisms that cause plant diseases are called plant _____.

5. The effectiveness of pesticides is diminished by _____ (two words), which is based on change in inherited traits.

6. The _____ (two words) causes pesticide use and expense to increase, while crop yields decrease.

7. _____ (two words) is the increase in concentration of a pesticide as it moves through a food web; the buildup of a pesticide within an organism's body over time is called _____.

8. The 1996 book entitled _____ (three words) warns that toxic chemicals are disrupting human hormone systems, causing cancers, birth defects, and lowered sperm counts.

9. A dust of the bacterium *Bacillus popilliae* can be applied to the ground as a biological control of _____ (two words) larvae.

10. Selective breeding is one technique used to develop pest-resistant crops. The process may take _____ to _____ years per crop variety.

11. *Bt* corn was one of the first _____ (two words) crops.

12. In 1998, *Bt* corn represented _____% of the U.S. corn crop.

13. The restriction of the importation of foreign pests and diseases (or exotic plant and animal materials that might harbor such problematic species) is called _____.

14. _____ (three words) combines biological, cultivation, and pesticide controls tailored to the crops and conditions of a particular farm.

15. Between 1985 and 1990, Indonesian _____ production decreased by 58%; during the same time period _____ production increased by 14%.

16. The 1954 _____ (two words) Amendment, also called the _____ Amendment, strengthened the Food, Drug, and Cosmetics Act (FDCA) of 1938 by establishing acceptable limits on pesticide levels in food.

17. First passed in 1947, the _____ (five words) Act (FIFRA) was designed to regulate pesticide effectiveness.

18. The amended FIFRA of 1988 did not address the issue of _____ contamination by pesticides and did not require pesticide companies to disclose the _____ ingredients in their products.

19. The _____ (three words) Act of 1996 establishes identical pesticide-residue limits for raw and processed foods; it also requires that the increased susceptibility of infants and children be considered in establishing these residue limits.

20. In the United States, the _____ (three words)
    Administration monitors toxic residues on imported fruits and vegetables but is able
    to inspect only about _____ % of food shipments.

21. In 1999, the Los Angeles Unified School District implemented the country's most
    aggressive policy to ban school use of _____.

22. _____ (two words) is the regular use of pesticides, regardless
    of pest population size; _____ is a technique based on monitoring
    pest populations so that pesticides are applied only when needed.

Short Answer:

1. If there were "perfect" pesticides, what characteristics would they have?

2. Name and give examples of the three major groups of insecticides.

3. What are some of the medical conditions the U.S. Department of Veterans Affairs
    recognizes as linked to dioxin exposure?

4. How do pesticides lead to declines in populations of pest predators? And why,
    despite a 33-fold increase in pesticide use in the United States since the 1940s, have
    crop losses to pests not dropped significantly?

5. Why is the incidence of pesticide poisoning highest in developing countries?

6. Why are children at greater risk from household pesticides than adults are, and what
    are the potential health effects?

7. Give an example of a biological control that has caused an unexpected problem, and
    name two disadvantages of the sterile male technique (a genetic control).

8. Why have some *Bt* crops begun to show diminished ability to deter pests?

9. How have insect pheromones been used to fight insect pests?

10. Why, despite its promising potential, have only a small proportion of U.S. farmers
    adopted IPM?

11. The EPA banned DDT use in 1972. What other pesticides have since been outlawed?

12. How are consumer attitudes linked to pesticide use?

13. What arguments are given for *not* banning all pesticides that cause cancer in
    laboratory animals?

Critical-Thinking Questions:

1. According to your text, "The short generation times… and large populations that are characteristic of most pests favor rapid evolution." Why?

2. Why would a pesticide ban increase food prices?

Data Interpretation:

1. If 2.7 million people died from malaria in 1999 (when world population reached 6 billion), what percentage of the world's population died from malaria that year?

2. What percentage of U.S. crops was lost to pests in 1974? 1989?

Chapter 23

Solid and Hazardous Wastes

Study Outline:

I. The Solid Waste Problem
   A. Types of Solid Waste

II. Disposal of Solid Waste
   A. Open Dumps
   B. Sanitary Landfills
   C. Incineration
   D. Composting

III. Waste Prevention
   A. Reducing the Amount of Waste: Source Reduction
   B. Reusing Products
   C. Recycling Materials
   D. The Fee-Per-Bag Approach

IV. Hazardous Waste
   A. Types of Hazardous Waste

V. Management of Hazardous Waste
   A. Chemical Accidents
   B. Current Management Policies
   C. Cleaning Up Existing Toxic Waste: The Superfund Program
   D. The Biological Treatment of Hazardous Contaminants
   E. Managing the Toxic Waste We Are Producing Now

VI. Environmental Justice
   A. Environmental Justice and Ethical Issues
   B. Mandating Environmental Justice at the Federal Level
   C. Environmental Justice and International Waste Management

VII. Integrated Waste Management

Key Terms: The following terms are listed in order of appearance in your textbook.

municipal solid waste
nonmunicipal solid waste
sanitary landfills
polymers
photodegradable
biodegradable
mass burn incinerators
modular incinerators
refuse-derived fuel incinerators
lime scrubbers
electrostatic precipitators
bottom ash (or slag)
fly ash
source reduction
Pollution Prevention Act
dematerialization
cullet
fee-per-bag approach
hazardous (or toxic) waste
dioxins
polychlorinated biphenyls (PCBs)
principle of inherent safety
Resource Conservation and Recovery Act (RCRA)
Comprehensive Environmental Response, Compensation, and Liability Act (CERCLA)
Superfund Act
Superfund National Priorities List
bioremediation
phytoremediation
environmental chemistry (or green chemistry)
environmental justice

Basel Convention
integrated waste management
product stewardship

Multiple Choice:

1. Representing a huge waste of high-quality plastics and metals, it is estimated that 150 million _____ will be disposed of in sanitary landfills by 2005.
    a. automobiles
    b. telephones
    c. computers
    d. VCRs
    e. golf carts

2. _____ produces more per-capita solid waste than any other country.
    a. Canada
    b. France
    c. Japan
    d. The United States
    e. Australia

3. Today's solid waste contains more _____ but less _____ than it did in the past.
    a. paper and glass, steel and plastics
    b. glass and steel, paper and plastics
    c. plastics and glass, paper and steel
    d. paper and steel, glass and plastics
    e. paper and plastics, glass and steel

4. _____ produces the most nonmunicipal solid waste.
    a. Mining
    b. Industry
    c. Health care
    d. Agriculture
    e. Forestry

5. Dumps and landfills typically produce _____ gas.
    a. methane
    b. ammonia
    c. hydrogen
    d. oxygen
    e. cyanide

6. The _____ requires owners to monitor a landfill for _____ years after it is closed.
    a. Environmental Protection Agency, 5
    b. Environmental Protection Agency, 30
    c. Bureau of Land Management, 5
    d. Bureau of Land Management, 30
    e. Fish and Wildlife Service, 50

7. Plastics are chemical *polymers*, composed of chains of repeating _____ compounds.
    a. carbon
    b. nitrogen
    c. phosphorus
    d. sulfur
    e. chlorine

8. One of the most difficult-to-manage solid wastes is _____, which, in vulcanized form, cannot be melted and reused.
    a. glass
    b. aluminum
    c. plastics
    d. steel
    e. rubber

9. Batteries, thermostats, and fluorescent lights should be removed from solid wastes prior to incineration to prevent _____ emissions.
    a. carbon dioxide
    b. sulfur dioxide
    c. nitrogen oxide
    d. mercury
    e. lead

10. Some electric utilities currently burn _____, which produce(s) as much heat as coal, often with less pollution.
    a. paper
    b. yard wastes
    c. tires
    d. sawdust
    e. plastics

11. Lime scrubbers and electrostatic precipitators are used to make _____ less polluting.
    a. paper production
    b. aluminum recycling
    c. incineration
    d. sanitary landfills
    e. polystyrene production

12. A useful way to deal with yard waste is to
    a. convert it to mulch and compost
    b. incinerate it to generate electricity
    c. incinerate it to generate steam for heating buildings
    d. dispose of it in sanitary landfills
    e. have homeowners burn it

13. The most underutilized aspect of waste management is
    a. dematerialization
    b. recycling
    c. reuse
    d. repair
    e. source reduction

14. The nation that recycles the highest percentage (27%) of its municipal solid waste is
    a. Japan
    b. the United States
    c. Sweden
    d. Canada
    e. Costa Rica

15. The U.S. is most successful (in terms of percent recycled) at recycling
    a. paper
    b. glass
    c. aluminum
    d. rubber
    e. plastic

16. Recycled, low-quality plastic mixtures can be used to make
    a. seat cushions
    b. beverage containers
    c. garden hoses
    d. "plastic lumber" for outdoor uses
    e. contact lenses

17. "Love Canal" is synonymous with
    a. rubber recycling
    b. radiation poisoning
    c. the glassphalt industry
    d. chemical pollution
    e. integrated waste management

18. Coal combustion, incineration of waste, metal recycling, copper smelting, and forest fires are among the known sources of
    a. radioactive wastes
    b. dioxins
    c. PCBs
    d. infectious wastes
    e. organic solvents

19. Which of the following is *not* a suspected health effect of dioxins?
    a. certain cancers
    b. endometriosis
    c. delayed fetal development
    d. decreased sperm counts
    e. liver damage

20. Which of the following is *not* a known effect of exposure to PCBs?
    a. harm to skin
    b. harm to eyes
    c. intellectual impairments
    d. liver and kidney damage
    e. heart disease

21. The Hanford Nuclear Reservation threatens soil, groundwater, and the Columbia River with _____.
    a. radioactive wastes
    b. toxic wastes
    c. infectious wastes
    d. all of the above
    e. a and b

22. The state with the greatest number of sites on the Superfund National Priorities List is
    a. Pennsylvania
    b. New Jersey
    c. California
    d. Texas
    e. New York

23. Phytoremediation is the use of _____ to remove toxic materials from soil.
    a. bacterial
    b. fungi
    c. plants
    d. single-celled algae
    e. PCBs

24. The most effective way to deal with hazardous waste is
    a. source reduction
    b. toxicity reduction
    c. incineration
    d. long-term storage
    e. discharge into sewers

25. The high incidence of _____ in many minority communities may be related to exposure to environmental pollutants.
    a. heart disease
    b. cancer
    c. tuberculosis
    d. asthma
    e. cataracts

26. The Basel Convention
    a. restricts international transport of hazardous waste
    b. seeks to educate U.S. citizens about toxic pollutants
    c. provides money for the cleanup of Superfund sites
    d. provides money for the construction of hazardous-waste landfills
    e. promotes international research on source reduction of hazardous waste

27. Which of the following is *not* gained from the industrial ecosystem of the J. R. Simplot Company, which produces McDonald's french fries?
    a. cattle feed
    b. methane for power plants
    c. diesel fuel
    d. fuel-grade ethanol
    e. irrigation water

28. The 17,000-acre Rocky Mountain Arsenal, a notorious Superfund site, is being cleaned up with the goal of becoming a
    a. new airport for the Denver area
    b. model community
    c. national wildlife refuge
    d. national park
    e. training facility for Superfund workers

29. New York City may partially alleviate its solid-waste problem by
    a. composting food wastes
    b. taxing residents for paper use
    c. exporting more trash to other states
    d. building new landfills in the metropolitan area
    e. dumping wastes into the sea

30. The goal of _____ is the basis of the Chrysler Corporation's Composite Concept Vehicle.
   a. outstanding gas mileage
   b. safety
   c. recyclable parts
   d. manufacturing efficiency
   e. a low-cost product

Matching: Match the phrases on the left with the responses on the right. Some responses are used more than once.

_____ 1. a component of many plastics; may release dioxins when incinerated

_____ 2. slower but much cheaper than other means of breaking down hazardous waste

_____ 3. trapped by air pollution control devices

_____ 4. poisoned hundreds in Japan and Taiwan

_____ 5. break down when exposed to sunlight

_____ 6. can be used to make glassphalt

_____ 7. makes up about 1% of the solid waste in the United States

_____ 8. hospital-waste incinerators probably produce the most

_____ 9. decomposed by microorganisms

_____10. linked to non-Hodgkin's lymphomas, among other human health problems

_____11. also called slag

_____12. byproducts of combustion of chlorine compounds

_____13. may contaminate meat, dairy products, and fish

a. photodegradable plastics
b. biodegradable plastics
c. polyvinyl chloride
d. bottom ash
e. fly ash
f. cullet
g. hazardous waste
h. dioxins

i. PCBs
j. bioremediation

Fill-In:

1. _____ (three words) consists of the solid items thrown away by homes, stores, offices, restaurants, schools, and other commercial and institutional facilities.

2. _____ (two words) receive more than half the solid waste generated in the United States.

3. The largest component of municipal solid waste is _____ (two words); the second largest is _____ (two words).

4. To prevent liquid waste from seeping into groundwater, landfills have layers of _____ (two words) and _____ (two words) at the bottom.

5. More than 100 landfills in the United States use _____ gas (produced by microorganisms decomposing organic material anaerobically) to generate _____.

6. The amount of _____ is growing faster than any other component of municipal solid waste.

7. The best materials to incinerate are _____, _____, and _____.

8. Some types of paper release _____ (a type of hazardous waste) when they are burned.

9. _____ incinerators are smaller than _____ incinerators, many of which are designed to recover the energy of combustion; _____ incinerators burn primarily plastic and paper, which has been shredded or shaped into pellets.

10. The _____ (two words) Act of 1990 was the first U.S. environmental law to focus on reduced production of pollutants at their place of origin.

11. The progressive decrease in the size and weight of products, through technological improvements, is called _____.

12. Dematerialization results in source reduction only if newer products are as _____ as the ones they replace.

13. Ninety-seven percent of the paper in _____ (what country) is recycled.

14. All recycling is driven by _____.

15. Charging households for each container of waste is called the _____ approach.

16. Incineration of medical and municipal wastes accounts for 70% to 95% of human-caused emissions of _____.

17. High-temperature incineration is one of the most effective ways to destroy _____.

18. Through an aspect of source reduction called the principle of _____ (two words), accidents are prevented because industrial processes are redesigned to involve less dangerous materials.

19. Passed in 1980, the Comprehensive Environmental Response, Compensation, and Liability Act (CERCLA) is commonly called the _____ Act; it established a program to clean up abandoned and illegal toxic waste sites in the United States.

20. The Superfund program spends much of its funding on _____ fees.

21. _____ chemistry seeks to redesign chemical processes to reduce environmental harm.

22. _____ (two words) is a fundamental human right to adequate protection from environmental hazards, regardless of age, race, income, gender, or other societal category.

23. _____ (three words) incorporates a variety of waste-reduction options into an overall waste management plan.

24. Those who adopt the concept of _____ (two words) seek to minimize their consumption of goods by buying less; instead, they may borrow or do without.

25. Manufacturers who embrace (or are forced to embrace) the concept of _____ (two words) assume responsibility for their products from cradle to grave.

Short Answer:

1. What are the major components of municipal solid waste?

2. What factors must be considered when landfill sites are chosen?

3. Why did the number of landfills in the United States decrease from 20,000 in 1978 to fewer than 2400 in 1999?

4. What problems are caused by tires in sanitary landfills?

5. In order of priority, what are the three goals of waste prevention?

6. Why might the cost of beverages increase if refillable glass bottles are used?

7. How did Toronto's city council and President Clinton promote paper recycling in the 1990s?

8. Name some uses of recycled tires.

9. What types of chemicals are considered hazardous? And, according to the EPA's Emergency Response Notification System database, which three states had the most toxic-chemical accidents in 1998?

10. What toxic chemicals can be removed or degraded by phytoremediation?

11. In order of their effectiveness, what are the three ways to manage hazardous waste?

Critical-Thinking Questions:

1. Using refillable glass bottles for beverages may lead to higher beverage prices. What costs *decrease* when refillable bottles are used?

2. Comment on the export of United States used paper to Mexico, China, Korea, and other foreign countries. In what ways does this export benefit the United States? What costs or disadvantages are there for the United States?

3. The text states that "[g]reater recycling occurs when the economy is strong than when there is a recession." Why?

Data Interpretation:

1. How many trees and gallons of water are saved when 30 tons of paper are recycled?

2. In the U.S., approximately 65 million people live in areas with curbside-collection recycling programs. What percentage of the U.S. population (based on the figure given in the *World Population Data Sheet*) is served by curbside-collection recycling?

PART SEVEN:

TOMORROW'S WORLD

Chapter 24

Tomorrow's World

Study Outline:

Multiple Choice:

1. Since the 1950s, humanity has cut down, without replacing, about _____ of the forests that existed then.
    a. one-tenth
    b. one-sixth
    c. one-fourth
    d. one-third
    e. one-half

2. It is estimated that _____ of all species have not yet been recognized and scientifically described.
    a. one-tenth
    b. one-fourth
    c. one-half
    d. two-thirds
    e. five-sixths

3. An estimated 65 million _____ die annually in the United States as a result of exposure to pesticides.
    a. fish
    b. birds
    c. beneficial insects
    d. frogs
    e. mammals

4. Generally, population growth rates are highest where there is
    a. the most extreme poverty
    b. the most open space
    c. the greatest wealth
    d. the highest level of education
    e. a low infant mortality rate

5. The rapidly growing population of Hyderabad, India, has reduced the cost of _____ by relying on residents' associations and reducing the city government's role.
    a. garbage collection
    b. public transportation
    c. public health
    d. maintaining parks
    e. construction of apartment buildings

Fill-In:

1. Although U.S. citizens make up only _____% of the world's people, they control approximately _____% of the world's economy.

2. About one in _____ people live in extreme poverty, and one in _____ are too undernourished for normal growth and bodily function.

3. During the second half of the 20th century, the world's human population grew from _____ billion to _____ billion.

4. Development that meets today's needs without compromising the ability of future generations to meet their needs is called _____ development.

5. Human cultural diversity is intertwined with _____ (two words).

6. The ultimate goal of _____ is to improve the quality of human life; this goal is complicated by the grossly uneven distribution of the world's _____.

7. Only about _____% of the world's scientists and engineers live in less developed nations, which, collectively, are the source of approximately _____% of the organisms humanity depends on.

8. _____ (two words), which protects topsoil, is an agricultural practice gaining popularity.

9. The carrying capacity of an ecosystem is determined by its ability to absorb _____ and renew itself.

10. To stay within Earth's carrying capacity, we must not only stabilize human population but greatly reduce excessive _____ and _____.

11. With minor exceptions, all that enters Earth is _____, and all that leaves is _____; thus there is no "away" place to put our wastes and pollutants.

12. Increasingly, knowledge and information are concentrated in _____.

13. The _____ (two words) in Rio de Janeiro in 1992 brought together more heads of state than any other meeting in history; however, there was little talk of _____ cooperation.

14. Most departments of environmental protection react to _____ instead of identifying _____ for their prevention.

15. At a deep, fundamental level, the most serious environmental problems are our own _____ and _____.

16. Humans' acquisition of wealth and material possessions is accomplished at tremendous _____ to the planet.

Short Answer:

1. What are some of the potentially renewable resources that humans have badly degraded in the past 200 years?

2. What are some of the ecosystem services humans rely on?

3. Explain the difficult and "paradoxical place" women hold in many societies.

4. What are some of the negative effects of agriculture that must be brought under control to ensure sustainable food production?

5. What are some of the attitudinal barriers to environmental protection and restoration?

6. What are some of the ways we can promote environmental awareness and action locally?

7. What did former West German Chancellor Willy Brant mean by "a blood transfusion from the sick to the healthy?"

8. How is it that we "contribute to a future in which neither our children nor our grandchildren will be able to live in anything like the affluence that we experience now?"

Critical-Thinking Questions:

1. How is it that so many people, especially those living in highly developed nations, have lost touch with nature and fail to see the disastrous consequences of abuse to the environment?

2. Remote employment, or telecommuting, has clear advantages; for example, it saves energy and time, and reduces pollution. Can you think of any environmental *harm* that can come from telecommuting?

Data Interpretation:

1. If telecommuting can reduce pollution by 1.8 million tons through a 15% reduction in transportation, how much can it reduce pollution through a 25% reduction in transportation?

ANSWERS

Chapter 1: Our Changing Environment

Multiple Choice:

1. a
2. b
3. e
4. b
5. c
6. c
7. b
8. d
9. d
10. b
11. b
12. c
13. e
14. a
15. c
16. d
17. b
18. b
19. c
20. d
21. a
22. e
23. b
24. e
25. c
26. a
27. a
28. c

Matching:

1. e

2. h
3. c
4. j
5. a
6. g
7. k
8. b
9. d
10. f
11. l
12. i

Fill-In:

1. Green architecture
2. Environmental science
3. 3
4. consumption
5. 50
6. estrogens
7. antagonism
8. endocrine
9. commercially extinct
10. tropical migrant
11. Fragmentation
12. Zebra mussels
13. exotic (or introduced, or invasive)
14. stratosphere, CFCs
15. ultraviolet
16. 21st
17. Carbon dioxide
18. climate change
19. watersheds
20. Ecology
21. coyotes

Short Answer:

1. We use nonrenewable fuels as if they were unlimited, we use renewable resources faster than they can be replenished naturally, we release toxins as if the environment's capacity to absorb them were limitless, and we allow our population to grow rapidly despite Earth's finite resources.

2. Alaskan and North Atlantic salmon, Peruvian anchovies, and Newfoundland cod of the Grand Banks are examples of fishery declines in recent years and decades.

3. Overfishing occurs because of increased demand for fish, increased numbers of fishing boats, and the use of high-tech methods, such as sonar and satellite images, to find fish. Also, there is intense international competition for dwindling fish populations.

4. Cerulean warblers, olive-sided flycatchers, yellow-billed cuckoos, eastern wood pewees, and wood thrushes are examples of declining tropical migrant songbirds.

5. The two major goals were to remove the gray wolf from the endangered species list and to restore the natural balance between predators and prey (especially by reducing oversized elk populations).

6. Comb jellies consumed so much plankton that anchovies and other fish species declined because of a diminished food supply. As fish catches have fallen off, the local Russian economy has suffered.

7. Major shifts in patterns of rainfall, melting ice sheets, rising sea levels, and coastal flooding are all expected results of global warming.

8. Critics of the Kyoto Protocol worry about the economic repercussions of stringent measures, such as a pollution tax, used to decrease greenhouse-gas emissions. Energy experts, on the other hand, argue that much of the required emission reductions can be met through energy conservation and increases in energy efficiency. For example, cars can be designed to get better gas mileage; also, more wind turbines, which generate electricity without releasing carbon dioxide, can be installed.

9. Giant trees, lianas (vines), orchids, ferns, and countless other plants died in the fire. Many animals lost their habitat and were forced to compete elsewhere (often in marginal habitats) for food and territories; among these were many species of neotropical songbirds.

10. Bio-prospecting involves investigating a variety of species as potential sources of useful and marketable chemicals, including drugs, flavorings, fragrances, and natural pesticides.

11. Issues discussed at the 1992 Earth Summit were (1) ways to combat global climate change, (2) ways to decrease the extinction rate (promote biodiversity), (3) the problem of deforestation, (4) Agenda 21 (an action plan of sustainable development), and (5) Earth Charter (a philosophical statement about environment and development).

Critical-Thinking Questions (Hints):

1. Think of all the foods and drinks you have consumed today. Are you wearing clothing that contains cotton or wool? Are you wearing leather? Have you used paper? Wooden furniture or building materials? Heat or electricity generated by burning fossil fuels? (Fossil fuels are the "fossils" of what organisms?) Where does atmospheric oxygen come from? Are you wearing perfume or cologne with a flower-based scent? Have you taken any herbal remedies? Did you pet or play with your dog or cat? And what organisms digest your sewage wastes?

2. How is this question related to the Endangered Species Act (ESA)? How did the ESA block the demands of ranchers? Do you feel this compromise is a valid one?

3. See the Envirobrief entitled "Welcome Home." What roles do beavers play in an ecosystem, and how might their numbers be affected by wolves?

Data Interpretation:

1. 1.3 billion/6 billion = .217 = 21.7%

2. After 1800, world population doubled once to reach 2 billion; it doubled once more to reach 4 billion, and it had achieved another half-doubling by 1999 to reach 6 billion. Therefore, world population doubled 2.5 times between 1800 and 1999.

3. $3000 \times .40 = 1200$ eggs hatching; $1200 \times .50 = 600$ survivors (20% of eggs laid)

Chapter 2: Addressing Environmental Problems

Multiple Choice:

1. e
2. b
3. a
4. b
5. d
6. d
7. b
8. b
9. d
10. e
11. c
12. c

13. c
14. a
15. b
16. b
17. d
18. e
19. c
20. c
21. d
22. b
23. a
24. d
25. a

Matching:

1. e
2. c
3. h
4. k
5. m
6. d
7. a
8. f
9. i
10. n
11. b
12. j
13. l
14. g

Fill-In:

1. scientific assessment
2. process
3. Scientific method
4. Inductive
5. Deductive
6. control, variable
7. Risk assessment
8. Risk management
9. Toxicology
10. more
11. lethal dose

12. effective dose
13. dose-response
14. mixtures
15. synergistic, antagonistic
16. more
17. ecological
18. Cost-benefit
19. cyanobacteria, algae, *Daphnia*
20. stewardship
21. nicotine

Short Answer:

1. Malaria is transmitted to humans by mosquitoes. Drainage pools along roads and small pools of water that collect in recently cleared areas of forests serve as mosquito breeding sites. Also, global warming allows malaria-transmitting mosquitoes to expand their ranges.

2. The five components of environmental problem solving are (1) scientific assessment, (2) risk analysis, (3) public education, (4) political action, and (5) follow-through.

3. *Repeatablility* means that when experiments and observations are repeated, they must produce consistent data.

4. The five basic steps of the scientific method are (1) identify a question or problem, and find out what is already known about it, (2) develop a hypothesis to answer the question or explain the problem, (3) design and perform an experiment to test the hypothesis, (4) analyze and interpret the data to draw conclusions, and (5) share new knowledge with the scientific community.

5. Rats and humans may respond differently to exposure to the same chemical. Also, rats are exposed to massive doses—relative to their body size—of suspected carcinogens; humans normally receive much smaller relative doses over longer periods of time. Therefore, extrapolating from rats to humans may not be scientifically sound.

6. Nutrients, especially nitrogen and phosphorus, stimulate the explosive growth of photosynthetic microorganisms, especially filamentous cyanobacteria. As these organisms die, deep-water bacteria that eat them multiply and consume great quantities of oxygen during the metabolism of decomposition. As oxygen levels drop, fish and many types of invertebrates begin to die.

7. The prediction was that continued addition of nutrient (i.e., sewage) pollution would result in a smelly lake—covered with mats of rotting cyanobacteria—unfit for swimming or drinking.

8. Puget Sound has a much greater quantity of water (for diluting pollution) than Lake Washington does. Also Puget Sound's natural abundance of nutrients means that cyanobacterial growth is not limited by phosphorus availability (if it were, populations would explode in response to added phosphorus). Instead, growth of cyanobacteria and algae in Puget Sound is limited by tides, which drive these organisms into deeper water where they cannot photosynthesize enough for rapid population growth.

9. Global commons are areas and resources that belong to no single individual, company, or country; therefore, they are available for exploitation by everyone. Examples are the atmosphere, fresh water, certain forests, wildlife, and the oceans' fisheries.

10. Poverty, overpopulation, and social injustice are some of the persistent problems linked to environmental problems.

Critical-Thinking Questions (Hints):

1. Recall that science is a *process* as well as a body of information. Also, science rests on a body of information that changes as new evidence comes to light. Does uncertainty imply absence of sound knowledge? Can you think of examples of uncertain information that is useful, even essential, to humanity?

2. Are there naturally occurring toxins and carcinogens in foods and water? How much of the risk we face, in terms of harm from food or water or air, comes from nature? How much comes from human-caused pollution? If we don't know, how can we find out? Is no-risk a reasonable and realistic goal for those who oversee the quality of air, water, and food?

3. What conflicting interests might politicians be trying to balance? What risks do they run by plunging ahead with demands for legislation when data are inconclusive or "experts" disagree on the best course of action? What ethical considerations might some politicians overlook and why? When are environmental and human health issues important enough to override economic and/or personal political considerations?

Data Interpretation:

1. 2.7 million/400 million = .00675 = .675%

2. 1 kg = 2.2 pounds; 17.5 mg/kg divided by 2.2 lbs/kg = 7.95 mg/lb
   140 lbs × 7.95 mg/lb = 1113 mg

Alternate approach: 140 lbs divided by 2.2 lbs/kg = 63.6 kg; 63.6 kg × 17.5 mg/kg = 1113 mg

Chapter 3: Addressing Environmental Problems, Part II

Multiple Choice:

1. d
2. e
3. a
4. a
5. b
6. c
7. b
8. d
9. a
10. a
11. a
12. b
13. a
14. b
15. e
16. d
17. b
18. e
19. c
20. c
21. b
22. b
23. a
24. c
25. c
26. e

Matching:

1. p
2. n
3. m
4. c
5. k
6. l

7. e
8. b
9. h
10. d
11. f
12. i
13. a
14. g
15. j
16. o

Fill-In:

1. unfunded mandates
2. Conservation
3. Henry David Thoreau
4. Muir
5. Civilian Conservation Corps
6. supply, demand
7. increases
8. environmental damage
9. goods
10. less
11. is not
12. low
13. Emission reduction credits (ERCs)
14. economic development
15. lead
16. decreased
17. old-growth
18. lowest
19. more
20. ethics
21. deep ecology worldview
22. net domestic product
23. economic degradation

Short Answer:

1. National forests are used for wildlife habitat, recreation, and timber harvest.

2. An EIS is an environmental impact statement. It must include (1) the nature of and need for the proposed federal action, (2) the long- and short-term environmental

effects of the proposed action, and (3) alternatives to the proposed action that will lessen adverse effects.

3. Federal agencies can no longer exploit resources on federally owned land (and nearly one-third of the land in the United States is federally owned) nor launch many public-works projects without environmental review.

4. Measuring the monetary cost of pollution is difficult, and the risks of environmental catastrophe are normally ignored.

5. Ideas for green taxes include taxes on leaded gasoline, taxes on active ingredients in pesticides, charges on nonreturnable containers, higher road tolls during rush hour, taxes on carbon dioxide releases, and taxes for private use of public water. The public objects to paying a tax on anything perceived as free, such as the right to use as much gasoline (and emit as much carbon dioxide) as desired.

6. The United States, the United Kingdom, Canada, and Japan have reduced sulfur-dioxide releases from coal-burning power plants.

7. Salvage logging involves the removal of dead, diseased, or damaged trees, along with "at-risk" healthy trees considered in danger of disease. Environmentalists weren't happy that the practice allowed loggers into areas previously off-limits to timber cutting. Also, both environmentalists and forestry scientists pointed out that dead and dying trees play important ecological roles prevented by their removal.

8. Measurement of the GDP fails to consider the economic costs of environmental degradation or the economic contributions of environmental protection and clean-up.

9. Signs of environmental devastation include severely polluted water; chemical spills; chemical haze of soot and sulfur dioxide; eroded buildings and statues; dead forests; declining crop yields; prevalence of human respiratory diseases; high rates of cancer, birth defects, and miscarriages; and low life expectancies.

10. High energy subsidies and lack of competition enabled power plants to provide energy at prices well below its actual cost.

Critical-Thinking Questions (Hints):

1. Missing from economists' equations are aesthetic, spiritual, psychological, educational, and ecological values (such as?) of ecosystems and their inhabitants. Also missing are the diffuse effects of globally distributed pollution problems (such as?). What makes such values and effects difficult to measure? Can you think of a better currency than money for making measurements of pollution costs? Is this currency globally applicable?

2. Must there always be winners and losers in environmental showdowns? Try to create a win-win scenario. How do you think the Northwest Forest Plan holds up as a *long-term* solution? Can you suggest improvements, or a different approach altogether?

3. How do the two worldviews differ in humanity's role and authority with respect to the environment? How do they differ in humanity's view of consumption? Of nature itself? What religious, social, economic, philosophical, and political underpinnings gave rise to the Western worldview? How will continued population growth, economic expansion, and rampant consumption of resources affect humans? Non-human species? Ecosystems?

Data Interpretation:

1. 160 billion board feet + 6 billion board feet = 166 billion board feet of white pine; 6 billion board feet/166 billion board feet = .036 = 3.6%

2. $170,000,000,000/268,000,000 people in the United States = $634/person living in the United States

3. 10.3 million tons – 5.5 million tons = 4.8 million tons; 4.8 million tons/10.3 million tons = .466 = 46.6%

Chapter 4: Ecosystems and Energy

Multiple Choice:

1. b
2. d
3. a
4. c
5. b
6. e
7. a
8. e
9. a
10. b
11. c
12. b
13. c
14. d
15. e
16. a

17. e
18. d
19. c
20. c
21. b
22. c
23. b
24. e
25. c
26. b

Matching:

1. g
2. e
3. n
4. d
5. l
6. a
7. o
8. p
9. f
10. m
11. b
12. b
13. k
14. k
15. j
16. j
17. q
18. q
19. c
20. h
21. i
22. i

Fill-In:

1. biotic, abiotic
2. Ecology
3. organization
4. Landscape ecology
5. Geology, earth science
6. kilojoules (kJ)

7. kilocalories (kcal)
8. potential, kinetic
9. closed, open
10. Entropy
11. $6O_2$, $6CO_2$
12. energy flow
13. detritus
14. food, oxygen
15. food chain
16. food web
17. trophic level
18. biomass
19. energy
20. gross primary productivity
21. net primary productivity

Short Answer:

1. Common insects in a Chesapeake Bay salt marsh are mosquitoes and horseflies; birds include seaside sparrows, laughing gulls, and clapper rails; invertebrates include shrimps, lobsters, crabs, barnacles, worms, clams, and snails. Saltwater dries out the skin of amphibians.

2. Energy occurs as chemical, radiant (or solar), heat, mechanical, nuclear, or electrical energy.

3. The first law of thermodynamics states that energy cannot be created or destroyed; it can only be transformed from one form to another.

4. The second law of thermodynamics states that whenever energy is converted from one form to another, some usable energy is converted to heat, a less usable form that disperses into the environment.

5. $6CO_2 + 12H_2O +$ radiant energy $\rightarrow C_6 H_{12}O_6 + 6H_2O + 6O_2$

6. Decomposers, parasites, and tree-dwelling herbivorous insects are typically more numerous than what they feed on. Pyramids of numbers do not give information about biomass at each trophic level or the amount of energy transferred from one trophic level to the next.

7. net primary productivity = gross primary productivity – plant respiration
       (plant growth)   (total photosynthesis)

8. The most productive aquatic ecosystems are algal beds, coral reefs, and estuaries. The open ocean is unproductive because it lacks available nutrients.

9. Bacteria form the base of food webs in hydrothermal vents. They support tube worms, clams, crabs, barnacles, and mussels. In deep water, sunlight is unavailable for photosynthesis.

10. Seals, penguins, and smaller baleen whales eat krill, which became more available to them as large baleen whales (which also eat krill) declined in number.

Critical-Thinking Questions (Hints):

1. What does the second law of thermodynamics say about entropy and heat? If all energy is ultimately converted to heat energy, what are the implications for life? For stars? For the universe as a whole?

2. What kinds of ecosystems have the greatest biodiversity? What happens to biodiversity as productivity increases? How does human use of fossil fuels and fertilizers affect the amount of nitrogen present in ecosystems? What impact, in terms of productivity, does nitrogen have on ecosystems? (See discussion of eutrophication in Chapter 2.)

Data Interpretation:

1. 250 kcal = 2500 kcal × .10 (.10 = 10%); therefore, approximately 2500 kcal of corn is required to produce 250 kcal of chicken meat.

2. The NPP for a temperate deciduous forest is listed as 1200 grams dry matter/m²/year. 650/1200 = .542 = 54.2%; in other words, agricultural land is 54.2% as productive as temperate deciduous forest.

3. If NPP = GPP – respiration, then GPP = NPP + respiration; therefore GPP = 17,034 kcal/m²/yr + 22,155 kcal/m²/yr = 39,189 kcal/m²/yr.

Chapter 5: Ecosystems and Living Organisms

Multiple Choice:

1. b
2. d
3. c
4. b
5. b
6. a

7. e
8. a
9. c
10. e
11. a
12. d
13. e
14. b
15. d
16. d
17. d
18. c
19. c
20. b
21. a
22. e
23. e
24. d
25. e
26. b
27. a
28. e
29. c
30. a

Matching:

1. e
2. b
3. k
4. m
5. f
6. g
7. l
8. c
9. h
10. d
11. j
12. a
13. i

Fill-In:

1. keystone species

2. coevolution
3. Nicotine
4. Symbiosis
5. symbionts
6. Zooxanthellae
7. Commensalism
8. Parasitism
9. pathogen
10. Competition
11. interspecific
12. habitat
13. fundamental niche
14. limiting factor
15. species diversity
16. ecotone
17. edge effect
18. community stability, ecosystem services
19. Evolution
20. natural selection
21. succession
22. pioneer community
23. Primary
24. Secondary

Short Answer:

1. It was hoped that the Nile perch would benefit fishermen and the local economy. The Nile perch first reduced populations of native fish called cichlids and then began to prey on freshwater shrimp.

2. The three main roles are producer (ex: red maple), consumer (ex: gray squirrel), and decomposer (ex: mushroom).

3. Various species of fig trees make fruits available throughout the year; therefore, when other fruits are scarce, monkeys, birds, bats, and other vertebrates, which normally don't favor figs, depend heavily on these fruits.

4. Having many eyes, ears, and noses watching, listening, and smelling for predators makes it less likely that a predator can approach the group and single out an individual without being detected.

5. Given time, one species would always thrive, while the other died out. One species always had a competitive advantage over the other; the victor was determined by the environmental (i.e., test tube) conditions Gause provided. This classic experiment demonstrates the competitive exclusion principle.

6. A complex community, composed of many species, provides many potential ecological niches. As organisms capable of filling those niches evolve, or migrate into the community, the community becomes even more complex, providing more ecological niches and continually promoting species diversity.

7. Species diversity is inversely related to a community's geographical isolation. An isolated area, such as an island or mountain peak, is likely to be small and to have fewer potential ecological niches than a larger area would have. Also, many species have difficulty reaching and colonizing a remote area, and local extinctions are likely because species are effectively "trapped" in their isolated locale when random events pose threats. And replacement of extinct species is slow because of the community's isolation.

8. Species diversity is reduced because a dominant species is able to appropriate a disproportionate share of resources. Other species are crowded out or out-competed, resulting in a community lacking diversity.

9. Darwin's four observations are (1) overproduction (the production of more offspring than will survive to maturity), (2) variation (the inheritable differences found in any population), (3) limits on population growth (diseases, predators, shortages of resources, etc., which curtail population growth), and (4) differential reproductive success (the greater survival rate of individuals who possess the most favorable combination of traits).

10. The typical sequence of succession is this: crabgrass→horseweed→broomsedge and other weeds→pine trees→hardwood (deciduous) trees.

11. Oaks have boom and bust years, in terms of acorn production. The boom years cause populations of white-footed mice to increase. These mice feed also on gypsy moth pupae, thereby helping the oaks, which are defoliated by gypsy moth caterpillars. In addition to white-footed mice, deer benefit from big acorn crops. They come to feed on the acorns, bringing deer ticks with them. The ticks' offspring feed on the mice, which often carry the bacterium that causes Lyme disease. Humans bitten by these ticks can be infected with the bacterium.

12. Declining fish populations have reduced the food supply for seals and sea lions. As populations of seals and sea lions have declined, killer whales (or orcas) have turned to sea otters as an alternative food source. Sea otter populations have been devastated as a result, and that's good news for sea urchins, the prey of sea otters. Burgeoning sea urchin populations, however, are devastating their prey: kelp forests.

13. The six kingdoms are Archaebacteria, Eubacteria, Protista, Fungi, Plantae, and Animalia.

Critical-Thinking Questions (Hints):

1. How could a competing species be helpful? (See the discussion of competition in this chapter.) What might happen if the competitor or predator is a keystone species? (See the discussion of Yellowstone's wolves in this chapter and in Chapter 1.) What benefits does a prey species gain from predation?

2. How might fish in the same lake or stream utilize resources differently? (Think of their spawning times and requirements, maturation rates, dietary options, etc.) How about plants? (Think of differences in root depths, flowering times, drought resistance, rate of growth, mineral requirements, etc.) Do you think two plants of different species, growing inches apart, can reduce competition by specializing in some way?

Data Interpretation:

1. $19,000/45,000 = .422 = 42.2\%$ as many native plant species in Canada and the continental United States as in Columbia, Ecuador, and Peru.

2. Assuming the original pair of elephants represents the first generation, 2^{10} elephants would be produced in 10 generations. (Use 2, not 4, because each *pair* of elephants, not each individual elephant, produces 4 offspring.) $2^{10} = 1024$ elephants.

Chapter 6: Ecosystems and the Physical Environment

Multiple Choice:

1. d
2. a
3. c
4. c
5. a
6. b
7. a
8. e
9. b
10. d
11. c
12. d
13. e
14. c
15. b

16. a
17. e
18. e
19. c
20. d
21. b
22. b
23. b
24. a
25. d
26. d
27. c
28. e
29. c
30. a
31. b
32. d

Matching:

1. i
2. m
3. e
4. e
5. e
6. e
7. g
8. k
9. b
10. l
11. c
12. a
13. d
14. h
15. f
16. n
17. j

Fill-In:

1. negative feedback
2. biogeochemical
3. closed
4. 0.03

5. combustion, oxidized
6. oceans
7. nitrogenase
8. nodules, legumes
9. nitrification
10. assimilation
11. ammonification
12. denitrification
13. Nitrogen, photochemical smog, acid deposition
14. phosphate
15. watershed
16. groundwater
17. stratosphere
18. mesosphere
19. thermosphere
20. gyres
21. La Niña
22. quickly, slowly
23. temperature, precipitation
24. rain shadow
25. tropical cyclones
26. storm surges
27. Mitch
28. boundaries, plate tectonics
29. Subduction
30. magma, lava
31. seismic waves, faults
32. rotation

Short Answer:

1. Marine phytoplankton remove carbon dioxide from the atmosphere by using it to make calcium carbonate shells. When they die, phytoplankton sink to the ocean floor and form seabed deposits, which ultimately become limestone. Removal of carbon dioxide from the atmosphere has prevented Earth from overheating.

2. The burning of fossil fuels releases carbon dioxide, as does the burning of tropical forests and the burning of wood for fuel. All these forms of combustion have increased since 1850, even more so since 1950.

3. The five steps of the nitrogen cycle are (1) nitrogen fixation, (2) nitrification, (3) assimilation, (4) ammonification, and (5) denitrification. Bacteria play a key role in all the steps except for assimilation.

4. Nitrogen enters bodies of water as a component of both commercial fertilizer runoff and acid deposition. It stimulates the growth of algae, some of which are toxic. When algae die, bacteria decompose them, consuming oxygen in the process. As a result, many fishes and other aquatic organisms may suffocate.

5. Seabirds defecate on land where they roost, and their wastes, called guano, contain large amounts of phosphorus. In this manner, some of the oceans' phosphorus returns to land, where it becomes available to plants.

6. Absorbed solar radiation runs the water (hydrologic) cycle, drives wind and ocean currents, powers photosynthesis, and warms the planet.

7. The sun's rays strike equatorial regions vertically, but they strike polar regions obliquely; in other words, the sun's rays are spread over a larger surface area near the poles. Also, their oblique angle means rays of light pass through more atmosphere near the poles than they do near the equator; hence, more of the sun's energy is scattered and reflected, which further contributes to cold temperatures.

8. The five layers of the atmosphere are troposphere, stratosphere, mesosphere, thermosphere, and exosphere.

9. Prevailing winds are the atmosphere's major surface winds, which blow more-or-less continuously. They are polar easterlies (near the poles), westerlies (in mid-latitudes), and trade winds (in the tropics).

10. The ocean conveyor belt is the global circulation of shallow and deep currents. When it reorganized 11,000 to 12,000 years ago, heat transfer to the North Atlantic stopped, resulting in extreme cold in North America and Europe, as well as overall global cooling.

11. The 1997–98 ENSO caused more than 2000 deaths and $33 billion in property losses. Heavy snows, ice storms, torrential rains, and droughts were all effects of the ENSO. The droughts in Indonesia exacerbated fires, resulting in destruction of vast areas of forest.

12. La Niña typically causes wetter-than-normal winters in the Pacific Northwest, warmer weather in the Southeast, droughts in the Southwest, and stronger and more numerous hurricanes in the Atlantic.

13. The six climate zones are tropical, dry, mild, continental, polar, and high elevation.

Critical-Thinking Questions (Hints):

1. On a day-to-day basis, does photosynthesis have a cooling or heating effect on climate? (What gas does photosynthesis remove from the atmosphere?) On the other

hand, what is the source of wood and fossil fuels, and what gas is released when they are burned?

2. How does the atmosphere change in density with increasing distance from Earth's surface? How could this density change affect temperature? Also, what layer of the atmosphere "contains" Earth's mountains, and how does its temperature change with increasing elevation?

Data Interpretation:

1. If 1 lb = 453.6 g, and 1 lb = 16 oz, 1 oz = 453.6 g per lb/16 oz per lb = 28.35 g. 4 oz = 28.35 g × 4 = 113.4 g; therefore, 12 g glucose/1 g nitrogen = x g glucose/113.4 g nitrogen; x = 12 g glucose/1 g nitrogen × 113.4 g nitrogen = 1360.8 g glucose.

2. Find the population (given in millions) of Guatemala on the *World Population Data Sheet*. Multiply this number by 22.4 kg/person. Multiply *this* number by 2.205 lb/kg to convert your answer to pounds.

3. If oceans reflect 5% of the sunlight hitting their surfaces, they *absorb* 95%. Oceans (and forests) have a very low albedo.

4. Each unit on the Richter scale corresponds to 30 times more energy. Therefore, a magnitude 8 earthquake releases 30 × 30 × 30 times more energy than a magnitude 5 earthquake. This equals 27,000 times more energy.

Chapter 7: Major Ecosystems of the World

Multiple Choice:

1. c
2. d
3. c
4. a
5. e
6. e
7. d
8. b
9. c
10. c
11. c

12. a
13. a
14. e
15. e
16. b
17. a
18. b
19. c
20. e
21. e
22. d
23. a
24. d
25. a
26. c
27. a
28. b
29. e

Matching I:

1. g
2. a
3. h
4. e
5. d
6. a
7. i
8. c
9. e
10. b
11. h
12. b
13. d
14. f
15. i

Matching II:

1. c
2. f
3. d
4. a
5. h
6. b

7. i
8. g
9. d
10. b
11. c
12. e
13. i
14. h
15. g

Fill-In:

1. biome
2. mediterranean
3. deserts
4. savanna
5. salinity
6. phytoplankton, zooplankton
7. Benthos
8. Standing, flowing
9. profundal, limnetic
10. thermocline
11. fall turnover, spring turnover
12. blooms
13. marsh, swamp
14. ecosystem services
15. Estuaries
16. intertidal zone
17. abyssal zone
18. zooxanthellae
19. fringing reef, barrier reef, atoll
20. neritic
21. oceanic
22. Everglades
23. bleaching, zooxanthellae (algae)

Short Answer:

1. Black-tailed prairie dogs' teeth (including enlarged incisors) and digestive tracts are modified for feeding on the seeds and leaves of prairie grasses, as well as insects, roots, and tubers. Prairie dogs live in large colonies, which enables them to alert one another to danger. They also burrow underground, creating networks of tunnels as safe living quarters. The piles of soil they excavate surround burrow openings, thereby helping prevent flooding during storms. In winter, prairie dogs' metabolisms

slow and they burn stored body fat; however, they become active on warm winter days and may come out of their burrows to feed.

2. Biomes are shaped by precipitation, average temperature, light, temperature extremes, rapid temperature changes, fires, floods, droughts, strong winds, and altitude.

3. U.S. grasslands provide (or once provided) habitat for bison, pronghorn elk, wolves, coyotes, prairie dogs, foxes, black-footed ferrets, various birds of prey, grouse, snakes, lizards, and many species of insects.

4. Humans have damaged desert vegetation and caused erosion by driving off-road vehicles (including army tanks) in deserts. Poaching of cacti and desert tortoises is another destructive human activity. Also, use of groundwater for homes, factories, and farms has caused aquifer depletion in certain deserts of the Southwest.

5. Savanna ecosystems are found in parts of Africa and South America. They are also found in western India and northern Australia.

6. The topmost story consists of scattered crowns of very tall (50 or more meters) broad-leaved evergreen trees; the middle story, which reaches a height of 30 to 40 meters, makes up a continuous canopy of leaves; the understory, which receives very little light, is populated by tree seedlings and herbaceous growth (ferns, mosses, orchids, bromeliads, etc.). Vegetation is dense only near streams or other openings in the canopy.

7. Aquatic ecosystems are strongly affected by salinity, amount of dissolved oxygen, availability of light, levels of essential mineral nutrients, temperature, pH, and the presence or absence of waves and currents.

8. Clearing of forests can cause erosion that smothers reefs under a layer of sediment. Also, increased salinity, caused by diversion of fresh water for human uses, can hurt coral reefs. Overfishing, water pollution, fishing with dynamite or cyanide, hurricane damage (possibly worsened by global warming), land reclamation, tourism, and the mining of corals for building materials are other human-related threats. In addition, divers and snorkelers cause damage by touching and kicking coral life, as well as by stirring up sediments, which can suffocate coral animals.

9. Coastal development and agriculture damage or destroy coastal ecosystems, including coral reefs and important spawning, feeding, and nursery areas; disease-causing organisms from human sewage contaminate shellfish and other organisms; trash entangles and kills marine animals; fertilizers, pesticides, heavy metals, synthetic industrial chemicals, oil, and ballast water, along with pollutants washed out of the atmosphere by precipitation, enter oceans and harm marine life; overfishing depletes fish populations; and shrimp- and scallop-harvesting techniques wipe out entire benthic communities.

10. To restore the Everglades, (1) farmers must clean up their runoff to reduce the amount of phosphorus entering the ecosystem, (2) at least 40,000 acres of farmland will be bought and converted to marshes that will purify agricultural runoff, and (3) the Army Corp of Engineers will re-engineer the area's system of canals, levees, and pumps to restore a more natural flow of water to the Everglades.

11. Examples of ecosystem services include air and water purification, waste decomposition, flood mitigation, crop pollination, pest control, soil renewal, and conservation of biodiversity. Ecosystem goods include food, timber, medicines, and fuels. A study of 16 biomes placed greatest value on wetlands, coastal and oceanic areas, and forests.

Critical-Thinking Questions (Hints):

1. Who do the world's oceans belong to? (See "Tragedy of the Commons" essay in Chapter 2.) Are the world's oceans connected to one another? What kind of legislation and enforcement is needed to safeguard the oceans, and what penalties might be effective deterrents for those tempted to ignore regulations?

2. What positive effect does fire have on soil fertility? What plants require fire and for what purpose? How might the disappearance of such plants affect other species? How can fire suppression alter an entire ecosystem, vastly changing its species composition? And, how can fire suppression lead to larger, more dangerous fires that threaten people as well as other organisms?

Data Interpretation:

1. 1 inch = 2.54 cm; therefore, 15 cm/2.54 cm per inch = 5.9 inches, and 350 cm/2.54 cm per inch = 137.8 inches.
137.8 inches – 5.9 inches = 131.9 more inches of rainfall per year in the tropical rain forest than in the desert.

2. 80 km × 160 km = 12,800 square kilometers (km^2).
1 km^2 = 0.4 mi^2; therefore, 12,800 km^2 × 0.4 mi^2/km^2 = 5120 mi^2.

Chapter 8: Understanding Population Growth

Multiple Choice:

1. e
2. b
3. c
4. d
5. b
6. c
7. a
8. a
9. d
10. b
11. a
12. d
13. d
14. b
15. e
16. e
17. c
18. c
19. c
20. d
21. c
22. a
23. b
24. d
25. d
26. e
27. a
28. a
29. b
30. b

Matching:

1. d
2. e
3. a
4. b
5. e
6. e
7. c

8. c
9. b
10. a

Fill-In:

1. Population ecology
2. Population density
3. natural increase
4. Immigration
5. biotic potential
6. life history characteristics
7. environmental resistance
8. carrying capacity
9. *r*
10. *K*
11. Survivorship
12. dependent
13. independent
14. 2.2
15. carrying capacity
16. Demographics
17. Replacement-level
18. total fertility
19. age structure
20. 32
21. Immigration Reform and Control
22. Utah, Florida; California, Texas, Florida

Short Answer:

1. Africa has a high incidence of sexually transmitted diseases, such as gonorrhea;
    infection with a sexually transmitted disease increases the likelihood of contracting
    AIDS if exposed to HIV. And, until recently, condom use has been rare.
    Urbanization, too, contributes to the spread of AIDS. Many job-seeking men move to
    cities, leaving their wives in the country. If they have sexual relationships with
    prostitutes, they put themselves at great risk of HIV infection.

2. A population's growth rate (*r*) is determined by birth rate (*b*), death rate (*d*),
    immigration (*i*), and emigration (*e*).

3. Environmental resistance includes a limited availability of food, water, shelter, and
    other essential resources, as well as limits imposed by disease (including those caused

by parasites) and predation. Environmental resistance increases as a population becomes crowded.

4. Organisms must balance their expenditure of energy between their own needs (such as growing and finding food), which promote their *individual* survival, and their need to reproduce, which ensures the *population's* survival. Too much energy expended on one of these two areas jeopardizes the other.

5. Tawny owls are *K* strategists. They regulate their reproduction so that it matches their food supply: Every year, some birds do not breed; if food availability drops, some birds don't incubate their eggs. Also, females rarely lay the maximum number of eggs, and they often delay breeding until late in the season when the rodents they eat are abundant.

6. Herring gulls have high mortality as chicks (Type III survivorship). Those that survive to adulthood, however, enjoy greatly increased survivorship, with the likelihood of dying remaining about the same thereafter (Type II survivorship).

7. Projections of human population growth depend on assumptions made about standard of living, resource consumption, technological innovations, and waste generation, all of which affect Earth's carrying capacity for humans.

8. Replacement-level fertility is always greater than 2.0 because some children die before they reach reproductive age. Developing countries have higher replacement-level values because of higher infant mortality rates.

9. Declining birth rates are linked to improved living standards (socioeconomic conditions). Other factors include urbanization, education of women, and increased availability of family planning services.

10. Population continues to grow in the U.S. because of (1) Baby Boom population momentum (more females having children) and because of (2) immigration.

11. Refugees face increasing violence in refugee camps, fees charged by human traffickers for illegal transport across borders, and refusal by many nations (both developing and highly developed) to accept them.

Critical-Thinking Questions (Hints):

1. What are the criteria that determine whether a species is *r*- or *K*-selected? Which pattern do you think the blue jay fits best? (See Table 8-1 in *Environment*, 3/e.)

2. What benefits and drawbacks do you think a quota system entails? Which groups are given immigration priority under the revised system? Do you think other groups also

deserve priority? If so, which ones? How would an open-door policy affect state and local governments? The environment of the U.S.? The global population issue?

3. How have wolves affected the size of the moose population? How did the moose population change after the decline of wolves in the 1980s? What affect did this change have on mountain ash and aspen trees? Why did nearly 2000 moose die between 1995 and 1997? Could a larger or smaller wolf population have prevented this die-off?

Data Interpretation:

1. $r = (b - d) + (i - e) = (12/1000 - 6/1000) + (1.5/1000 - 0.8/1000)$
    $= (0.012 - 0.006) + (0.0015 - 0.0008) = 0.006 + 0.0007 = .0067 = 0.67\%$ per year

2. $0.009 = (8/1000 - 4/1000) + (i - 0.7/1000)$; $0.009 = (0.008 - 0.004) + (i - 0.0007)$;
    $0.009 = 0.004 + (i - 0.0007)$
    subtracting 0.004 from both sides gives $0.005 = i - 0.0007$; adding 0.0007 to both sides gives $i = 0.0057$

3. 2029 – 1999 (which is the year human population reached 6 billion) = 30 years.
    30 years/12 years per 1 billion = 2.5 billion; 2.5 billion + 6 billion = 8.5 billion

Chapter 9: Facing the Problems of Overpopulation

Multiple Choice:

1. e
2. b
3. c
4. d
5. c
6. e
7. e
8. d
9. a
10. a
11. b
12. d
13. a
14. a
15. b
16. c

17. d
18. c
19. e
20. d
21. e
22. c
23. b
24. d
25. a
26. e
27. a
28. c
29. b

Matching:

1. e
2. a
3. h
4. f
5. i
6. b
7. g
8. j
9. c
10. d

Fill-In:

1. developing
2. 300
3. environmental degradation, poverty
4. economic, population
5. sustainable
6. natural resources
7. overpopulated
8. Consumption, people
9. more
10. Population, Organization, Environment, Technology
11. brownfields
12. urban, developing
13. Compact development
14. developing
15. faster

16. total fertility rate
17. Smaller
18. Population Activities
19. 1.25 billion (1,250,000,000); 1.8
20. high unemployment
21. lower
22. higher

Short Answer:

1. The three approaches to solving the world hunger problem are based on different beliefs about the main causes of hunger. One approach focuses on reducing population growth. Another promotes economic development—including technological advances that increase food production—in developing countries. A third approach favors efforts to distribute resources more equitably.

2. Huge debts for past development projects preclude future loans, thereby impeding further economic development.

3. The two generalizations are these: (1) Although the natural resources required for an individual's survival are small, rapid population growth can overwhelm and deplete a country's resources, and (2) in highly developed nations, individual resource demands are large (far beyond what survival requires), causing pollution and resource depletion.

4. Poor people often must use natural resources in unwise, unsustainable ways for the purpose of immediate survival. Such exploitation of resources diminishes long-term prospects of economic development, however, and contributes to ongoing poverty.

5. $I = P \times A \times T$ (environmental **i**mpact = number of **p**eople × **a**ffluence per person [a measure of consumption] × environmental effects of **t**echnologies used)

6. Urban populations have grown and rural populations have shrunk because (1) increased mechanization allows fewer farmers to support more people and reduces job opportunities in rural places, and (2) cities traditionally provide many jobs through industry, education, government, and other opportunities.

7. Environmental problems associated with urbanization include destruction of wildlife habitat through urban/suburban sprawl, air pollution from auto and industrial emissions, water and soil pollution from urban runoff, and the urban-heat-island effect. Environmental benefits include—at least potentially—efficient use of land; energy and pollution savings through use of public transportation systems; and (with compact development) reduced need for highways and parking lots, which frees land for more beneficial uses.

8. Cities in many developing nations are growing so fast that services and jobs lag behind population growth. As a result, substandard housing, extreme poverty, high unemployment, severe pollution, and inadequate water supplies and sewage- and waste-disposal services are the norm.

9. Despite the encouraging-sounding percentage change, the actual number of women who are not using family planning services has risen because of population growth.

10. Laws determine many aspects of a country's culture that affect fertility rates. These aspects include minimum age of legal marriage; amount of compulsory education; a tax structure that encourages or discourages large families; and availability of funds for family-planning services, education, health care, old-age security, and incentives for smaller or larger families.

11. In India, economic development (e.g., adult literacy) and family planning (e.g., population education programs) have been linked. Also, multi-media ads and education promote birth control, while improving health services lower infant- and child-mortality rates. Other efforts are aimed at making contraceptives more available, improving women's status and increasing birth spacing.

12. The U.N. Population Fund expects underfunding of the World Programme of Action to cause over 100 million unintended pregnancies (resulting in many millions of births and abortions), tens of thousands of maternal deaths, and millions of infant and child deaths.

Critical-Thinking Questions (Hints):

1. How does affluence harm the environment and exhaust natural resources? How might energy and manufacturing technologies become more environmentally friendly? How might residents of developed nations change their habits and lifestyles to lessen their impact on the environment?

2. Think of improvements that could be made in transportation, housing, land use, recycling and waste disposal, wildlife habitat, recreation opportunities, aesthetics, air and water quality, etc. Brazil's city of Curitiba may serve as a springboard for ideas.

3. Do you see overpopulation as a pressing global problem? What lessons have been learned from China's and India's efforts to reduce population growth? Do your religious beliefs support family planning? If you expect to have a family, how will you decide how many children to have?

Data Interpretation:

1. environmental impact = number of people × affluence per person (or consumption per
    person) × technological impact;
    environmental impact = 150,000,000 people × 13,000 miles/year × .03 g/mile =
    58,500,000,000 g/year = 58,500,000 kg/year

2. secondary education: $0.4 \times 2.8 = 1.12$
    primary education: $0.4 \times 4.7 = 1.88$
    no formal education: $0.2 \times 5.5 = 1.10$
    total fertility rate (TFR) = $1.12 + 1.88 + 1.10 = 4.1$

3. $1.8/5.8 = .31 = 31\%$

Chapter 10: Fossil Fuels

Multiple Choice:

1. b
2. d
3. a
4. e
5. c
6. a
7. b
8. c
9. a
10. a
11. d
12. d
13. e
14. a
15. c
16. d
17. e
18. c
19. d
20. b
21. b
22. a
23. e
24. c
25. b

26. a
27. d
28. b
29. e
30. d

Matching:

1. h
2. d
3. e
4. k
5. a
6. f
7. l
8. c
9. j
10. m
11. b
12. n
13. i
14. g

Fill-In:

1. oil
2. Organization of Petroleum Exporting Countries
3. Strategic Petroleum Reserve
4. 8
5. oil, natural gas
6. Coal
7. electricity
8. bituminous coal
9. anthracite
10. coal
11. dragline
12. carbon dioxide
13. coal
14. acids
15. 5.6, acidic
16. Petroleum or crude oil
17. Natural gas
18. continental shelves
19. Russia

20. Los Angeles
21. oil
22. 19
23. gas hydrates
24. China
25. subsidies, do not

Short Answer:

1. Increased prices for gasoline and home heating oil, long lines at filling stations, lowered car sales, increased popularity of fuel-efficient foreign cars, and reduced speed limits were some of the effects.

2. Increased gas consumption in the past two decades is the result of relatively inexpensive oil, which makes gasoline relatively cheap and readily available. There are more vehicles on the road than ever, partly because of increased population. Also, U.S. citizens are driving larger, more powerful vehicles, which, aggravated by higher speed limits, consume more fuel per mile.

3. The two biggest uses of energy in the United States are highway vehicles and the heating, ventilation, and air conditioning of buildings.

4. SMCRA stands for the Surface Mining Control and Reclamation Act (passed in 1977), which requires restoration of surface-mined lands. Funds have been cut, thereby limiting inspection and enforcement, and reducing the likelihood that many areas mined and abandoned before 1977 will ever be restored.

5. Surface mines cause water pollution through the drainage of acids and toxins from abandoned sites. Also, sediment from soil erosion and landslides pollutes local waters.

6. Atmospheric carbon dioxide prevents heat from escaping into space. Global warming causes polar ice caps to melt, which can lead to higher sea levels and coastal flooding. Coastal erosion and the loss of coastal buildings are related threats. (Other effects of global warming are discussed in Chapter 20.)

7. Oil and natural gas are easier than coal to transport; also, they burn more cleanly.

8. It is difficult to predict how long oil and natural gas supplies will last because we do not know how much more oil and natural gas will be discovered. We also do not know what technological breakthroughs might increase the amount of fuel extracted at each deposit. Nor do we know if future consumption will be higher or lower than— or the same as—today's.

9. Natural gas, when burned, releases fewer hydrocarbons and much less carbon dioxide than other fossil fuels do; it also contains almost no sulfur, which contributes to acid deposition. As an auto fuel, natural gas releases *much* less toxic material, carbon monoxide, and soot than gasoline does. In addition, natural gas is abundant in North America; therefore, it has the potential to reduce U.S. dependence on foreign oil.

10. Birds (ducks, loons, cormorants, bald eagles, etc.), sea otters, and killer whales were among the victims of the Alaskan oil spill.

11. Proponents of oil exploration in the Arctic National Wildlife Refuge say development of domestic oil will improve the balance of trade and make us less dependent on foreign oil. And, the location is near Prudhoe Bay, which is declining in its own oil production but has the infrastructure to store and transport oil from the refuge. Proponents argue further that there will be little lasting impact on the environment and wildlife. Those *opposed* to oil exploration argue that a very temporary oil supply does not justify permanent threats to a fragile wilderness ecosystem, and they point out signs of significant environmental damage, some of which will never be repaired, at Prudhoe Bay. Also, why not use the money for research into renewable energy sources and energy conservation instead? And won't present draining of U.S. oil result in greater future dependence on foreign oil?

12. Objective 1: Increase energy efficiency and conservation.
Objective 2: Secure future fossil fuel energy supplies.
Objective 3: Develop alternative energy sources.
Objective 4: Accomplish the first three objectives without further damaging the environment.

13. To reduce our use of gasoline, we can avoid unnecessary trips, carpool, ride a bicycle, ride a bus or train, walk (which means living close to work, school, and retail areas), drive in a more fuel-efficient way, keep our cars well-tuned, keep our tires inflated at the recommended pressure, remove unneeded weight from our cars, and buy cars that are fuel efficient.

Critical-Thinking Questions (Hints):

1. How might we more efficiently heat and cool our homes and workplaces? What changes in agriculture and the transportation of food would save energy? Do we need all the products we buy? If not, what can we do without? How can transportation needs be met in more fuel-efficient ways? What incentives might promote energy conservation?

2. What factors influence the present rate of coal consumption? (Think of cost, availability, human population growth, the ultimate decline of oil reserves, economic development in China, India, etc.) What future technologies could decrease our dependence on fossil fuels? Who gains if coal use increases? Who loses?

3. Think of economic, environmental, and national security issues surrounding the import of oil. Also, do you see any correlation with U.S. employment opportunities?

4. What will the U.S. do once it has exhausted its *own* oil reserves? What about the U.S. trade deficit, which grows with our increasing dependence on *foreign* oil? Consider also environmental costs of the transport and extraction of oil.

5. Can the original topography of the land be precisely duplicated? How long will it take for the forest vegetation to grow back? When replanting is done, are all the original species returned to the site? Does anyone know what those species were? What effects can loss of biodiversity have?

Data Interpretation:

1. 14,000 BTU per pound/7000 BTU per pound = 2.0 = 200%

2. 6530 BTUs = energy output per person per mile (car)
   939 BTUs = energy output per person per mile (bus)
   6530 BTUs/939 BTUs = 6.95 times more BTUs for the automobile driver

3. $2.00 per gallon – $1.30 per gallon = $.70 per gallon

4. If a car requires one gallon of gas to go 36 miles at 55 mph, it will require 1.5 gallons (50% more) to go 36 miles at 75 mph. One gallon is two-thirds of 1.5 gallons; therefore, the car will be able to travel only 24 miles (two-thirds of 36 miles) on 1 gallon. The answer is 24 mpg.

Chapter 11: Nuclear Energy

Multiple Choice:

1. e
2. b
3. a
4. a
5. d
6. c
7. b
8. c
9. e
10. c

Answers

11. a
12. a
13. c
14. b
15. c
16. e
17. d
18. b
19. d
20. a
21. c
22. b
23. b
24. e
25. d
26. e
27. b
28. b
29. b

Matching:

1. g
2. j
3. e
4. k
5. a
6. h
7. c
8. m
9. b
10. l
11. i
12. d
13. n
14. f

Fill-In:

1. half-life
2. nonrenewable
3. fuel rods
4. neutrons
5. breeder

6. meltdown
7. 170,000
8. Uranium-235, plutonium-239
9. mutations
10. oncogenes
11. nuclear warheads
12. plutonium
13. Low-Level Radioactive Policy
14. a volcano, active earthquake fault lines
15. groundwater
16. Strontium-90, cesium-137, krypton-85
17. 40
18. Shippingport
19. fusion
20. Standardizing
21. less

Short Answer:

1. The Manhattan Project was the top-secret U.S. project to develop a nuclear bomb during WWII. Einstein recognized the relationship between atomic mass and energy (expressed as $E = mc^2$) that allowed scientists to grasp the potential of nuclear weapons. Also, when he fled the Nazis and emigrated to the U.S., Einstein told the President (FDR) that the Germans were trying to develop a nuclear bomb; this information initiated the Manhattan Project.

2. A nuclear reaction releases 100,000 times more energy than a chemical reaction does.

3. In 30 years, 50% of the radioactive material has decayed; in the next 30 years (total time elapsed: 60 years), half of what remains decays, leaving 25% still radioactive. In 30 more years (total time elapsed: 90 years), half of the remaining 25% decays, leaving 12.5% radioactive; in 30 more years (total time elapsed: 120 years), 6.25% remains, and so on.

4. In breeder fission, U-238 is converted to plutonium (P-239); this isotope of plutonium is fissionable material that can be used to make nuclear weapons.

5. Nuclear power plants are very costly to build because of their large size, the many years required to plan and build them, and the slowness and complexity of the regulatory process, which causes construction to fall behind schedule.

6. Deregulation gave customers choices about where to buy their electricity, thereby causing competition among utility companies. Some nuclear power plants (generally the older ones) that are not cost-competitive with other sources of electricity have been closed, and others are likely to follow suit.

7. The Three Mile Island accident was the result of human error after a mechanical (valve) failure. The Chernobyl accident was the result of both human error (at least partly because of inadequate training) and flawed design of the nuclear reactor.

8. Contaminated breast milk prevents mothers from nursing their babies. Also, in some areas, water is unsafe to drink and local milk, meat, and produce is unsafe to eat. In addition, many thousands of people have had to relocate to safer locations, and the local economy has suffered from loss of agricultural production.

9. Some areas were affected while others were spared. Overall, the distribution was uneven and unpredictable. (See Figure 11-8 in *Environment*, 3/e.)

10. Health effects (to date) include death, birth defects, mental retardation in newborns, leukemia in infants, thyroid cancer and immune abnormalities in children, and psychological injuries.

11. Ionizing radiation can damage DNA. Changes (mutations) in DNA can convert normal genes to oncogenes, which cause cells to divide abnormally and create cancerous tumors.

12. Options for storing nuclear wastes include underground rock formations; aboveground, thick-walled mausoleums; Antarctic ice sheets; and locations beneath the ocean floor. Most experts favor underground storage in stable rock formations.

Critical-Thinking Questions (Hints):

1. What forms of pollution are generated by both types of power? What are the human costs? Also, think about security and economic considerations.

2. What are some of the threats to safe, long-term storage of nuclear wastes? How might these threats be dealt with? How can information be transmitted without relying on words that might not be understood in the distant future?

3. What can computers do better than humans? What can humans do better than computers? How can people and computers act synergistically together? Who is responsible for the consequences if a computer malfunctions?

Data Interpretation:

1. If it takes 6600 years for 50% of the plutonium waste to decay, it will take another 6600 years for half of the remaining 50% to decay. This means 6600×2, or 13,200 years, will pass before 75% of the waste has decayed.

2. x kg/500 metric tons = 0.45 kg/7300 metric tons;
    x kg/500 metric tons = 0.0000616 kg per metric ton;
    x = 0.0000616 kg per metric ton × 500 metric tons = 0.0308 kg uranium;
    1 kg = 2.205 lb, and 2.205 lb per kg × 0.0308 kg = 0.0679 lb;
    0.0679 lb × 16 oz per lb = 1.09 oz uranium

Chapter 12: Renewable Energy and Conservation

Multiple Choice:

1. e
2. c
3. e
4. a
5. c
6. b
7. a
8. b
9. d
10. a
11. d
12. b
13. b
14. e
15. c
16. c
17. d
18. b
19. a
20. b
21. a
22. d
23. e
24. e
25. e
26. c
27. a
28. c
29. a

Matching:

1. j
2. d
3. b
4. f
5. i
6. k
7. a
8. g
9. e
10. c
11. d
12. h
13. h
14. c
15. f

Fill-In:

1. wind farms
2. infrared radiation
3. Photovoltaic solar cells
4. fuel cell
5. biomass
6. biogas, biogas digesters
7. methanol, ethanol
8. cow manure
9. Denmark
10. Schistosomiasis
11. Wild and Scenic Rivers
12. ocean waves
13. ocean thermal energy conversion
14. Geothermal energy
15. hot, dry rocks
16. tides
17. Energy conservation
18. Energy efficiency
19. Energy intensity
20. National Appliance Energy Conservation
21. Cogeneration
22. energy audits
23. Net metering

Short Answer:

1. Electricity cannot be generated at night or on cloudy days. As a backup, natural gas is often used to generate electricity at such times.

2. A solar power tower is a solar thermal electric system with a tall tower surrounded by computer-controlled mirrors that track the sun. The mirrors focus solar radiation on a receiver at the top of the tower, thereby heating molten salt, which produces steam for electric generation.

3. *Advantages*: Utility companies can purchase PVs in small modular units, at modest expense, to increase generating capacity. Also, PVs are more economical than power lines in remote areas, PV costs continue to decline, and PVs generate electricity with minimal maintenance and no pollution, even in cloudy weather.
    Disadvantages: PVs are only about 10% efficient at converting light energy to electricity, large-scale electrical production requires a great deal of land for solar panels, and electricity produced by PVs is not yet economically competitive with other sources of electricity.

4. Biomass production requires land and water that might otherwise be used for food crops, and, if food crop production decreases, food prices go up. In addition, the use of wood as fuel can lead to deforestation, desertification, soil erosion, air pollution, and degraded water supplies. If crop and forestry residues are used as biomass, the soil does not gain the erosion-preventing, nutrient-enriching benefits these residues provide when left on the land. Also, massive plantings of biomass crops reduce biological diversity and wildlife habitat, and burning biomass releases carbon dioxide.

5. *Advantages*: Costs are declining, making wind power the most cost-competitive of all forms of solar energy; generating electricity through wind power is non-polluting; and combining wind farms with raising crops or cattle is a profitable and productive use of land.
    Disadvantages: Wind turbines can kill migratory birds and detract from the natural beauty of the landscape. Also, many areas are not windy enough for profitable use of wind power.

6. The filling of a large reservoir behind a dam can cause earthquakes.

7. GHPs utilize the temperature difference that exists between Earth's surface and subsurface. The ground (subsurface) provides heat in winter and coolness in summer; GHPs are used to efficiently cool and heat buildings.

8. Energy conservation and efficiency save energy for future use, buy time for energy research and development, cost less than development of new energy sources and supplies, stimulate the economy by generating new technologies and business opportunities, and reduce environmental damage caused by energy production and consumption.

9. Minivans, sport utility vehicles, and light trucks have become popular; they have lower average gas mileages than sedan-type cars.

10. Energy use can be reduced by replacing furnaces and appliances with more efficient models and by installing storm windows, storm doors, and thicker attic insulation. Adding heat pumps, caulking cracks around windows and doors, and insulating air-conditioning ducts also reduce energy use. Other measures apply to both old and new homes: shading the house with trees, installing window shades and awnings, using ceiling fans (to reduce the need for air conditioning) and fluorescent bulbs, lowering temperature settings on water heaters, and installing low-flow shower heads and faucets that reduce hot-water use.

11. We can lower our thermostat settings in winter and raise them in summer, turn off lights when we leave a room, drive more slowly (to get better gas mileage), use carpools and public transportation, and put pressure on elected officials to enact and enforce energy-conservation measures.

Critical-Thinking Questions (Hints):

1. What is a quick and easy way to obtain a list of threatened and endangered freshwater species? Is it relevant that most of these species do not have economic importance? What ecological roles might they play that promote the health of freshwater ecosystems? Is it possible to change existing dams and water flows to restore downstream ecosystems without losing the dams' hydroelectric roles? (Information on the Grand Canyon ecosystem may provide insights.) What should be done when a dam is "silted in," no longer able to provide hydroelectric power?

2. How is most electricity generated in developed countries (which use the most electricity)? In other words, what is burned? What does "burning" release that is related to global climate change?

3. What are the major ways that electricity is used at your school? How about fossil fuels? Do you know how dorms and classroom buildings are heated? Are you aware of any energy-conserving practices or technologies already in use? For example, are trees used to shade buildings in summer? Are buildings designed to take advantage of the winter sun? Is use of buses, bicycles, and carpools encouraged? Are students directly charged for energy use in their residences? Can you think of some energy-saving incentives?

4. What ways of producing energy are most and least destructive to the environment? Do you think energy conservation and energy efficiency can balance population growth? Can you come up with win-win energy scenarios, those with economic *and* environmental benefits?

Data Interpretation:

1. $90 − $4 = $86; $86/$90 = .956 = 95.6\%$ drop in price

2. 1.5 pounds/kWh × 30,000 kWh = 45,000 lbs;
    1 kg = 2.205 lbs; therefore, 45,000 lbs = 45,000 lbs/2.205 lbs/kg = 20,408.2 kg
    carbon dioxide

3. 7630 megawatts − 6115 megawatts = 1515 megawatts

Chapter 13: Water: A Fragile Resource

Multiple Choice:

1. d
2. a
3. d
4. d
5. b
6. a
7. a
8. e
9. d
10. c
11. b
12. e
13. c
14. e
15. c
16. b
17. b
18. a
19. a
20. d
21. c
22. e
23. c
24. c
25. b
26. b
27. b
28. a
29. e

Answers

30. d

Matching:

1. g
2. k
3. n
4. d
5. i
6. l
7. m
8. e
9. a
10. j
11. f
12. h
13. b
14. c

Fill-In:

1. reservoir
2. vaporizes, sublimates
3. hydrologic cycle
4. watershed
5. confined (or artesian) aquifer
6. semiarid
7. increases
8. 40
9. aquifer depletion
10. salinization
11. Reclaimed
12. Stable
13. Agency for International Development (AID)
14. Kazakstan
15. Groundwater
16. Sustainable water use
17. Glen Canyon
18. smolts
19. Distillation, desalinization
20. Microirrigation
21. xeriscaped

Short Answer:

1. The San Francisco Bay area is heavily developed; therefore urban and suburban runoff, as well as boat traffic, add pollutants to the water. Also, much of the water that once flowed into the Bay is diverted for irrigation, industries, and drinking water; reduced freshwater flow causes more saltwater to intrude from the ocean. Declining water quality has caused several fish species to become endangered; meanwhile, exotic species multiply at the expense of native aquatic species.

2. Freshwater use is increasing because human population is increasing; per capita use is increasing, too.

3. In nature, water contains dissolved gases (such as carbon dioxide and oxygen) as well as dissolved mineral salts. These normal "additions" do not make water unclean or unsafe.

4. Most groundwater is considered nonrenewable because it takes hundreds to thousands of years for water to accumulate in an aquifer; therefore, it is easily withdrawn for human uses faster than it is recharged by percolation.

5. South America's Amazon River and Africa's Congo River have the largest watersheds, measuring 6144 thousand square kilometers and 3807 thousand square kilometers, respectively.

6. Smaller levees farther from a river's edge are less expensive to build, result in less flood damage, and allow some of the natural benefits of floods—such as improved wildlife habitat and replenished floodplain soils—to occur.

7. Draining wetlands, building on floodplains, and constructing levees all worsened the effects of the flooding of the Mississippi in 1993. Wetlands are important as natural sponges that moderate floods; developed floodplains provide numerous structures for water to damage; paved surfaces reduce the land's natural ability to absorb water; and levees cause floodwaters upstream to surge, resulting in worse flooding in areas without levees.

8. Water consumption in the West and Southwest averages 44% of renewable water; elsewhere in the United States, consumption averages 4% of renewable water.

9. Both bodies of water have suffered from the diversion of freshwater (for human needs) that once flowed into them. Because both Mono Lake and the Aral Sea have no outlet, evaporation is the only natural outflow; evaporation removes only water, not dissolved salts, which accumulate and make the decreasing volume of water increasingly salty. Wildlife declines as a result. Also, dust-and-salt storms occur when wind lifts exposed soil off the former lakebed; these storms threaten human health.

10. India has a long dry season and a shorter wet season. Most precipitation that falls (heavily) during the wet season drains away into flooded rivers; therefore, total runoff is high, but stable runoff—that which is available throughout the year—is low.

11. States may require permits to drill wells, limit the number of wells in a given area, and restrict the amount of water pumped from each well.

12. The Edwards Dam on the Kennebec River prevented 10 migratory fish species from going upstream to spawn. Dams on the Sandy River damaged the natural habitat of two threatened species: chinook salmon and steelhead trout.

13. The downstream reaches of the Missouri require adequate water flow for navigation, irrigation, electrical power, and domestic consumption. But the upstream regions require water for a multimillion-dollar fishing and tourism industry. Also, farmers and environmentalist disagree on the building of additional dikes and levees, which protect crops from floods but also reduce natural habitats. Meanwhile, Native Americans want freedom to use the Missouri's water in a variety of ways, including for irrigation and hydroelectric power.

14. Industries involved in (1) chemical products, (2) paper and pulp, (3) petroleum and coal, (4) primary metals, and (5) food processing use nearly 90% of industrial water in the United States.

15. Water conservation is promoted by installing water meters and charging by amount consumed instead of using a flat fee. Another approach is to offer rebates to homeowners who install water-saving devices, such as low-flush toilets. Also, building codes can be changed to require installation of water-conservation fixtures (water-saving showerheads, faucets, and toilets). Leaky pipes and water mains can be repaired, and the price of water can be increased to reflect its actual cost. Furthermore, xeriscaping (see Envirobrief in *Environment*, 3/e) can be promoted as an alternative to watered lawns.

16. Despite population growth, water use declined in the United States between 1980 and 1995 because of increased use of water-efficient appliances, reduced industrial water use, new irrigation techniques that reduce evaporation, and increased public awareness of the need for water conservation.

Critical-Thinking Questions (Hints):

1. See Figure 13-7 *b* and *c* in *Environment*, 3/e. How does a natural, non-channelized river slow the speed of water? How would water speed differ in a channelized stream? What happens when the water hits a downstream area that is not channelized?

2. Why does erosion increase when land is cleared for farming? How can livestock cause reduced plant cover, and how is plant cover related to erosion? Is agricultural damage to soil more likely in an arid or semiarid area than in a wetter region? If so, why? If crops are inadequate for people's needs, what do they do to ward off hunger?

3. How does microirrigation differ from conventional flood irrigation? (See Figure 13-25 in *Environment*, 3/e.) Why does less evaporation occur, and how is evaporation related to the accumulation of salts?

4. Does your scenario involve legislation, negotiation, compromise, fund raising, media campaigning, scientific research, sustainable resource use, humanitarian sensitivity, environmental protection, etc.? Good luck!

Data Interpretation:

1. 68.3% for irrigation/8.6% for domestic/municipal = 7.9 times more water used for irrigation than for domestic/municipal uses

2. 7.5 million acre-feet × 8 people per acre-foot = 60 million people

3. Of the 2.5% of the world's water that is fresh, 0.5% is groundwater; 0.5%/2.5% = .2 = 20% (of the world's freshwater is groundwater)

Chapter 14: Soils and Their Preservation

Multiple Choice:

1. b
2. a
3. c
4. e
5. b
6. c
7. d
8. b
9. d
10. e
11. e
12. a
13. b
14. a
15. a

Answers

16. c
17. e
18. d
19. e
20. b
21. c
22. c
23. c
24. c
25. a
26. a
27. b
28. a
29. d

Matching:

1. d
2. g
3. l
4. m
5. a
6. h
7. n
8. j
9. b
10. e
11. o
12. c
13. k
14. i
15. f

Fill-In:

1. Desertification
2. 99
3. carbon, oxygen
4. carbonic acid
5. Australia, South America, India
6. profile, horizons
7. Castings
8. mycorrhizae, mycelia
9. Nutrient cycling

10. sand, silt, clay
11. acidic
12. spodosols
13. Alfisols
14. Mollisols
15. aridisols
16. oxisol
17. Agroforestry
18. no-tillage
19. strip cropping
20. terracing
21. organic
22. compost
23. Conservation Reserve

Short Answer:

1. Soil provides water, a medium in which plants anchor roots, and thirteen of the sixteen minerals essential for plant growth (see Table 14-1 in *Environment*, 3/e).

2. The components of soil are (1) mineral particles, (2) organic matter, (3) water, and (4) air.

3. Everglades soils have an unusually high organic-matter content, composed of partly decomposed sawgrass. When such soils are exposed to air, through the draining and plowing that accompany agriculture, decomposition of the organic matter accelerates, and the soil level subsides.

4. Ants aerate the soil by making tunnels, add organic matter by carrying food to their nests, and plant seeds by burying them.

5. The two organisms are vascular plants and fungi. The fungi absorb soil minerals and transfer them to plant roots, thereby enhancing plant growth; the plants provide food, produced through photosynthesis, for the fungi.

6. Thousands of variations in climate, vegetation, parent material, underlying geology, topography, soil organisms, and soil age result in thousands of different soil types, which vary in color, depth, mineral content, pH, aeration, drainage, etc.

7. Essential minerals and organic matter wash away when soil erodes. Fertilizer is applied to help restore the soil's lost fertility.

Answers

8. Reducing soil erosion reduces the flow of sediments into water. Sediments can cloud the water, decrease photosynthesis, and smother bottom-dwelling organisms. Reducing soil erosion also reduces pollution of water by pesticides and fertilizers, which wash off the land with soil.

9. Leaves and stems shield the soil and cushion the impact of raindrops; also, roots hold the soil in place.

10. Most of the mineral nutrients in a tropical rainforest are stored in the vegetation, not the soil, which is nutrient poor. When trees are cut for timber, or burned, their stored nutrients are removed from the ecosystem. Then crops quickly exhaust the soil's meager reserves of nutrients, leaving the soil infertile.

11. Organic fertilizers are longer lasting; in other words, they don't quickly leach away, as inorganic fertilizers do. Because they are less soluble, organic fertilizers are less likely to cause water pollution. Also, organic fertilizers improve the soil's water-holding capacity and sometimes suppress soil organisms that cause plant diseases. Other advantages of organic fertilizers are that they do not produce the air pollutants (nitrous and nitric oxides) associated with inorganic fertilizers and do not require the great energy expenditure that goes into inorganic-fertilizer production.

Critical-Thinking Questions (Hints):

1. What are the constituents of loam? What benefits are provided by the different sizes of mineral particles? How would these benefits promote plant growth?

2. What was growing in the Dust Bowl region 100 years ago? How do those plants differ from wheat in their response to drought? What would be the effect of this difference on the soil in a time of severe drought?

Data Interpretation:

1. 6.6 billion metric tons per year (India) + 5.5 billion metric tons per year (China) = 12.1 billion metric tons per year;
    12.1 billion metric tons per year/2.24 billion people = 5.4 billion metric tons per billion people per year

2. 76 million hectares of conservation tillage/43 million hectares of conventional plowing = 1.77 times more hectares of conservation tillage

3. 7.7 metric tons per hectare – 0.6 metric tons per hectare = 7.1 metric tons per hectare;
    7.1 metric tons per hectare/7.7 metric tons per hectare = .922 = 92.2%;
    7.1 metric tons saved per hectare per year × 10 years = 71.0 metric tons (of soil) saved per hectare

Chapter 15: Minerals: A Nonrenewable Resource

Multiple Choice:

1. b
2. d
3. e
4. e
5. c
6. b
7. c
8. d
9. e
10. a
11. c
12. d
13. e
14. b
15. b
16. b
17. c
18. a
19. b
20. e
21. c
22. d
23. c
24. a
25. d
26. a
27. e
28. d

Matching:

1. h
2. d
3. i
4. b
5. a
6. f
7. c
8. j
9. l

10. g
11. k
12. e

Fill-In:

1. opposes
2. Minerals
3. iron
4. Rocks
5. Chile
6. sedimentation
7. diamonds
8. Smelting
9. Acid mine drainage
10. sulfur
11. derelict
12. funding
13. wetland
14. hyperaccumulator
15. chromium
16. reserves
17. resources
18. life index
19. manganese nodules
20. Reuse
21. low-waste
22. sustainable manufacturing
23. does not

Short Answer:

1. The original purpose of the law was to encourage settlement in the West. The law has no provisions for environmental protection; hence, mining companies have left billions of dollars worth of destruction.

2. All gold mining methods produce enormous amounts of waste; an extraction method called *cyanide heap leaching* leaves behind vast volumes of *toxic* waste, which can threaten waterfowl, fishes, and drinking water supplies. Other methods cause erosion (which clogs streams and threatens aquatic life), acid mine drainage, and heavy-metal contamination. Finally, open-pit gold mining involves pumping great quantities of water out of the ground, an action that can lower the water table and cause springs to dry up.

3. Techniques include aerial or satellite photography, measurements of gravity and the Earth's magnetic field, sampling at sites geologists identify based on their knowledge of the crust's formation, seismographs, and computer analysis of ocean-depth data.

4. An advantage of subsurface mining is that less land is disturbed or destroyed; disadvantages include greater expense and risks to miners.

5. Surface mining destroys vast areas of natural topography and habitat. Loss of vegetation results in erosion; wind and water erosion pollute air and waterways; and acid mine drainage pollutes soil and water, often devastating aquatic ecosystems and contaminating groundwater.

6. Sulfur, cadmium, lead, arsenic, and zinc, found in many ores, can pollute the air during the smelting process.

7. No. The United States has less than 5% of the world's population, yet it consumes roughly 20% of many metals. (See Figure 15-7 in *Environment*, 3/e.)

8. The five top mineral producers are the (1) United States, (2) Canada, (3) Australia, (4) the Russian Federation, (5) and South Africa.

9. Future quantities depend on many unknown factors: new discoveries of mineral deposits, the development of substitutes, rates of consumption (based partly on economics), the cost of energy, new technologies for extraction, political pressures, and environmental costs.

10. The U.N. Convention on the Law of the Sea is an international treaty that views the minerals and organisms of the open sea as belonging to all humans. It requires the establishment of an international group to oversee seabed mining and sell mining rights; the money collected is earmarked as aid to developing countries.

11. Gold, lead, nickel, steel, copper, silver, zinc, and aluminum are recycled. Recycling extends life indices of minerals, saves land from destruction, reduces the need for solid waste disposal, and reduces energy consumption and pollution. It is also a key step in changing attitudes, helping to convert throwaway societies into low-waste societies.

Critical-Thinking Questions (Hints):

1. Where are U.S. coal mines (vs. mineral mines) located, relative to population centers? Compared to the mining of minerals, how much coal is mined, and how much acreage is consumed by strip mining? And who pays for reclamation? Who pays for clean-up of Superfund sites?

2. From what raw material are plastics made? Is this a renewable or nonrenewable resource?

Data Interpretation:

1. $15,000,000,000/52 sites = $288,461,530/site

2. 6,000,000,000 people × .046 = 276,000,000 (276 million) people in the United States.

3. 1 gallon equals 128 ounces. 128 ounces/6 ounces per can recycled = 21 1/3 cans recycled

Chapter 16: Preserving Earth's Biological Diversity

Multiple Choice:

1. e
2. a
3. e
4. c
5. b
6. a
7. d
8. c
9. d
10. b
11. b
12. c
13. e
14. d
15. a
16. a
17. c
18. d
19. c
20. e
21. b
22. d
23. c
24. b
25. b
26. a

27. a
28. e
29. b
30. e

Matching:

1. j
2. h
3. o
4. d
5. s
6. b
7. l
8. c
9. g
10. p
11. m
12. u
13. n
14. e
15. a
16. q
17. v
18. k
19. f
20. i
21. r
22. t

Fill-In:

1. have not
2. Ecosystem services
3. extinct
4. background extinction
5. mass extinction
6. endangered, threatened
7. Florida, California, Hawaii; Hawaii
8. tropical rain forests
9. adaptive radiation
10. 35
11. Islands
12. Commercial harvest

13. Conservation biology
14. In situ, ex situ
15. Seed banks
16. national conservation strategy
17. population, community
18. flyways
19. commercial extinction
20. Japan
21. 1
22. Wildlife ranching, game farming

Short Answer:

1. Forests provide lumber and maintain the purity of fresh water. They also help control flooding, prevent soil erosion, and provide habitat for many species. Insects pollinate flowering plants. Animals, fungi, and microorganisms interact in ways that keep populations of various species in balance. They also develop and maintain fertile soils. Bacteria and fungi are responsible for most decomposition, which enables nutrients to cycle through ecosystems.

2. Popular, high-yield varieties tend to be genetically uniform, which makes them susceptible to disease and pests. These crop varieties can be crossed with ancestral varieties to make them more resistant to damage. Genetic engineering, too, can be used to introduce genes that confer protection.

3. Cherry and horehound (cough medicines); periwinkle, mayapple, tunicates, red algae, mollusks, corals, and sponges (anti-cancer drugs); certain beetles (steroids with birth-control potential); fireflies (an antiviral compound); and centipedes (a fungicide) all provide pharmaceutical compounds. Various plants provided oils, lubricants, fragrances, dyes, paper, lumber, waxes, rubber and other latexes, resins, poisons, cork, and fibers. Various animals provide wool, silk, fur, leather, lubricants, and waxes.

4. Species are at greater risk of extinction if they require large territories, live on islands, or have very small ranges, low reproductive success, specialized breeding areas, or specialized feeding habits. Small population sizes and lack of genetic diversity are other threats to long-term survival.

5. Species on islands often have small populations that are not supplemented by immigration and are unable to emigrate. Often, they have evolved in isolation from competitors, predators, and disease organisms; therefore they have few defenses when such organisms are introduced by humans.

6. Humans damage habitats by building roads, parking lots, bridges, and buildings; by clearing and logging forests; by draining marshes; by mining minerals (including

fossil fuels); and by using natural resources for recreation (golfing, skiing, camping, "off-roading," etc.).

7. Humans are responsible for habitat loss, exotic species (biotic pollution), pollution, and overexploitation.

8. Wild animals are removed from their breeding populations, shrinking their populations' gene pools. Also, the capture of adults may result in the death of their young. Many animals die in transit, while waiting to be sold, or in their new captive environments; these losses are replaced with more wild animals. And, the high prices people are willing to pay for "exotic" pets promotes illegal harvesting and supports a lucrative black market.

9. Habitat loss, pollution, increased UV radiation, infectious disease, and global climate change have all been offered as explanations for amphibian declines.

10. Especially in developing countries, inadequate money and expertise for resource management, along with lax enforcement of conservation laws, are limitations. Another limitation is protection of areas—such as tundra, deserts, and high-altitude mountains—which have great scenic beauty but relatively low biological diversity. Many ecosystems with great biological diversity, particularly in the tropics, are underprotected.

11. Some plants have seeds that are intolerant of being dried out and thus cannot be stored. Also, seeds cannot remain alive indefinitely and must be periodically germinated to produce new seeds; this process is expensive. Fires and power failures can wipe out genetic diversity; therefore, seed samples must be subdivided and stored in several different seed banks. Finally, plants stored as seeds cannot evolve in response to environmental changes, and their successful reintroduction to nature may consequently be undermined.

12. Some opponents of the ESA say it is too costly; others say it infringes on property rights, fails to provide economic incentives for private landowners, and impedes economic progress. Conversely, other critics believe ESA is underfunded and less effective than it should be.

13. CITES is the Convention on International Trade in Endangered Species of Wild Flora and Fauna. Its chief goal is to protect endangered and threatened species harmed by the highly lucrative international wildlife trade. Unfortunately, CITES' effectiveness is limited by lax and erratic enforcement and mild penalties.

14. To reverse the extinction trend we need to increase public awareness of the importance of biological diversity, support conservation-biology research and the establishment of an international system of parks, control pollution (especially those pollutants that contribute to acid rain, global warming, and ozone depletion), and encourage conservation efforts in by providing economic incentives. Examples of

economic incentives include ensuring that countries gain financially from the use of native genetic resources (including new drugs), promoting ecotourism, and forgiving or reducing debts in exchange for protection of biological diversity.

15. Game animals cause less erosion (than cattle do) because they eat a greater variety of plants and do not permanently damage vegetation. Also, game animals require less water, thereby lessening the amount diverted from nature for agricultural needs.

Critical-Thinking Questions (Hints):

1. What have you eaten (including spices) and drunk in the past 24 hours? Have you worn clothing made from animal or plant products? Have you used furniture made of wood? Perfume from flowers? Paper? What else? (This list may be surprisingly long.)

2. Some thinkers tie the "masters of the planet" view to the origins of agriculture. How could farming change humans' relationship with nature? Which religious traditions have the most ancient roots? Do any predate agriculture? Do 19th- and 20th-century descriptions of hunter-gatherers shed any light on these questions?

3. If you agree, how can you respond to the claim that people matter more than other species and will increasingly require the 10% for agriculture and other human activities? If you disagree, how can you respond to the argument that vital ecosystem services must continue if humanity is to survive? And that all species have a role to play and a right to exist? If land is to be preserved in a natural state, what selection criteria are most important? Presence of rare and endangered species? Unusual communities? Pronounced biodiversity? Scenic beauty? Contribution to human welfare?

Data Interpretation:

1. Table 16-1 gives a total of 1880 imperiled species. The plant total is 1055; $1055/1880 = 0.561 = 56.1\%$
    The invertebrate total is 331; $331/1880 = 0.176 = 17.6\%$
    The mammal total is 85; $85/1880 = 0.045 = 4.5\%$

2. $2000 - 1999 = 1$; $1/2000 = 0.0005 = 0.05\%$

Chapter 17: Land Resources and Conservation

Multiple Choice:

1. d
2. c
3. e
4. a
5. d
6. b
7. e
8. d
9. c
10. b
11. b
12. c
13. d
14. e
15. c
16. b
17. b
18. e
19. e
20. a
21. a
22. b
23. c
24. a
25. d
26. c
27. d
28. c
29. d
30. e

Matching:

1. e
2. j
3. o
4. c
5. h
6. g
7. n

Answers

8. a
9. m
10. d
11. k
12. b
13. l
14. i
15. f

Fill-In:

1. Wild and Scenic Rivers
2. Land and Water Conservation Fund
3. natural regulation
4. carbon dioxide
5. Ecologically sustainable
6. clearcutting
7. deforestation
8. more
9. 97
10. population growth
11. Southeast Asia
12. Latin America
13. boreal
14. the South
15. temperate rain
16. carrying capacity
17. desertification
18. horses, burros
19. Wetlands
20. Wetlands Reserve Program
21. National Wetlands Coalition
22. Cameroon, biodiversity
23. National Environmental Research Parks

Short Answer:

1. Natural areas help prevent pest outbreaks (by providing pest predators), floods, and erosion. They also recharge groundwater, break down pollutants, recycle wastes, and promote biodiversity. *And*, they provide opportunities for education, research, recreation, solitude, and psychological replenishment.

2. Most federally owned land is managed by the Bureau of Land Management (BLM), the Fish and Wildlife Service (FWS), the National Park Service (NPS), and the U.S. Forest Service (USFS).

3. Additions to the National Wilderness Preservation System are normally opposed by businesses that utilize public lands: timber, mining, ranching, and energy companies.

4. Grizzly bears probably need more room than the parks provide. (They have vast ranges.) Also, grizzlies and humans don't mix well; some bears are destroyed when they become nuisances or threats to humans. (And who knows what health problems bears develop from eating human garbage?)

5. From trees and forests we obtain wood for fuel, construction, furniture, paper products, etc. We also obtain nuts, fruits, mushrooms, and medicines, not to mention employment for millions. Forests cool the local climate and promote rain by providing moisture for clouds. They also remove carbon dioxide from the atmosphere (which helps combat global warming) and play important roles in the carbon, nitrogen, and other biogeochemical cycles. Trees and forests release oxygen into the atmosphere, reduce erosion and mudslides, and regulate the flow of water (thereby diminishing the impact of droughts and floods). Finally, forests maintain biodiversity by providing essential habitat for countless species, and they offer humans spiritual sustenance and recreation.

6. Forests absorb, hold, and slowly release water. They form humus-rich, water-holding soils, a protection in times of drought. They also promote the formation of clouds, which can ease drought by providing rain. In times of flood, forests absorb water and slow its movement, thereby reducing flood damage.

7. Deforestation leads to soil erosion. Streams carry silt to reservoirs, where it builds up behind dams, potentially interfering with the flow of water needed for hydroelectric power generation.

8. Deforestation is caused primarily by (1) subsistence agriculture, (2) commercial logging, and (3) cattle ranching.

9. Temperate forests have expanded in parts of the East because of secondary succession on abandoned farms, commercial planting, and government protection.

10. Road building in national forests is controversial because taxpayers fund roads that allow private logging companies to profit from federally owned forests. Also, road building is environmentally destructive, accelerating erosion and mudslides, damaging stream banks, polluting aquatic habitats, fragmenting wildlife habitat, and providing routes of entry for disease organisms and exotic species.

11. Rangeland management includes seeding, building fences to prevent overgrazing, burning shrubs, suppressing weeds, and protecting biological habitats.

Answers

12. Wetlands provide food and habitat for waterbirds and other wildlife, including many endangered species; help reduce flood damage; improve water quality by trapping nutrients and other pollutants; provide humans with wild rice, blackberries, cranberries, blueberries, and peat moss; and offer many recreational and educational opportunities.

13. In 1992, Congress asked the NRC to settle an issue, namely the definition of wetlands, that had become highly controversial because of resistance by farmers and real estate developers to enforcement of wetland protection. Because most wetlands are privately owned, property-rights advocates have strongly resisted legislation that seeks to protect and restore wetlands.

14. Prior to a land-use decision, a thorough inventory (soil type, topography, species diversity, endangered or threatened species, historical or archaeological sites) should be completed. The current value of the land, including ecosystem services, should also be appraised.

Critical-Thinking Questions (Hints):

1. What kinds of stresses put pressure on wildlife populations? How might these be harder for small, isolated populations to deal with? Why would a larger population be more likely to recover from, say, a 75% reduction?

2. How might hunting and fishing affect local economies? What groups might pressure politicians to allow hunting and fishing in wildlife refuges? Historically, what type of wildlife has been "protected" by most national wildlife refuges? In what ways has the world changed (since Teddy Roosevelt's time) that might affect the hunting/fishing issue?

3. What, besides houses, must be constructed when a new development is built? What services and conveniences do people expect to find close to home? Imagine that you are the planner of a new community. What can you do to conserve open space and minimize road construction, parking lots, etc., without denying people access to the goods and services they need?

4. See the *World Population Data Sheet* in the back of your text. If it does not give 1998 data, use "Natural Increase" information to estimate the U.S. population in 1998.

Data Interpretation:

1. $1.35 per month per cow × 12 months = $16.20 per year per cow; $13,500,000 per year/$16.20 per year per cow = 833,333 cows

2. 89.4 million original hectares – 42 million remaining hectares = 47.4 million hectares lost; 47.4 million hectares/89.4 million hectares = .53 = 53% lost
    104 acres/42 hectares = 2.476 acres/hectare (Officially, 2.471 acres = 1 hectare; the values given for acres and hectares in the text have been rounded off, resulting in a slight discrepancy.)

Chapter 18: Food Resources: A Challenge for Agriculture

Multiple Choice:

1. d
2. d
3. e
4. e
5. b
6. a
7. c
8. c
9. d
10. e
11. e
12. a
13. b
14. d
15. c
16. e
17. e
18. a
19. c
20. c
21. e
22. d
23. b
24. d
25. b
26. d
27. a
28. e
29. e
30. e

Answers

Matching I:

1. a
2. b
3. c
4. a
5. d
6. e
7. c
8. d
9. b
10. e

Matching II:

1. b
2. a
3. f
4. b
5. e
6. b
7. c
8. b
9. e
10. d

Fill-In:

1. Organic Food Production, 3
2. more
3. undernourished
4. malnourished
5. 3 (approximately 50%)
6. marasmus
7. undernutrition/malnutrition
8. 70, 1993
9. 250,000 kilocalories (10% of what is eaten)
10. yield
11. Germplasm
12. green revolution
13. 40
14. food additives, preservatives, antioxidants, coloring agents
15. Food and Drug Administration (FDA)
16. Latin America and sub-Saharan Africa
17. genetic engineering

18. bycatch
19. Open management, ocean enclosure
20. Magnuson-Stevens Fishery Conservation and Management
21. Aquaculture, mariculture
22. essential amino acids

Short Answer:

1. Overnutrition causes obesity, high blood pressure, and an increased risk of diabetes and heart disease. It may also increase the risk of certain cancers.

2. Severe famines have struck Africa, especially the nations of Ethiopia, Sudan, and Somalia. More recently, North Korea has experienced famine.

3. A new disease or introduced pest could wipe out millions of acres of genetically uniform crops, causing widespread famine. Use of genetically diverse crops (some of which would likely have natural resistance to a new disease or pest) and reliance on many species would help protect humanity from such a disaster.

4. In developing countries, the green revolution has promoted dependence on imported fossil fuels and mechanized equipment, at the expense of traditional agriculture. The high-energy costs and heavy pesticide use of the green revolution are related to environmental costs: air and water pollution. Also, important food sources—such as millet and sweet potatoes, which grow well in parts of Africa—have been overlooked by green-revolution genetic tinkering. Subsistence farmers typically do *not* benefit because they cannot afford the substantial outlays of energy and capital required by green-revolution agriculture.

5. Indiscriminate use of antibiotics leads to the development of antibiotic resistance in bacteria. This problem is a growing threat to human health.

6. Many pests have developed or are developing resistance to pesticides; this forces farmers to apply increasing amounts of these toxic chemicals. Pesticide residues contaminate foods and kill beneficial soil organisms. In addition, pesticide runoff can kill aquatic life in lakes, rivers, and estuaries.

7. Irrigation has become increasingly costly, underground water sources (aquifers) are being depleted, soils rendered too salty by the evaporative residue of irrigation water are continuously being abandoned, and water is increasingly diverted for other uses (residential and industrial).

8. Heavy mulching lessens soil infertility and erosion. Insect damage is reduced by planting several crops together, and the use of a legume (such as a bean) improves soil fertility by adding nitrogen.

9. The GM soybean contained a gene from Brazil nuts, and people allergic to Brazil nuts were allergic to the GM soybean, too.

10. Listed as "at risk" are haddock, Atlantic cod, Peruvian anchovy, Cape hake, southern bluefin tuna, capelin, chub mackerel, and Japanese pilchard. Resources of the open ocean are not claimed by individual nations; therefore, they are subject to exploitation—largely without regulation—by countless seafood companies, research interests, sportsmen, etc. from many, many ports. Pollution of the seas is another largely unregulated threat. The increasing pressure on fisheries stems from the growing human population's need for protein and from technological advances in fishing gear that have increased fishing efficiency.

11. Overfishing will be reduced by establishing quotas for various species, restricting use of certain types of fishing gear, limiting the number of fishing boats, and closing fisheries during spawning periods.

12. Coastal mangroves are cut down to make room for shrimp farms. These mangroves (which grow in water) provide breeding areas for many fishes. There are also potential water pollution issues associated with shrimp farming.

Critical-Thinking Questions (Hints):

1. What, besides availability of animal products, determines people's food choices? Think of cultural, religious, and income-level differences that relate to diet. How might climate play a role? Politics? Marketing?

2. Is all the world's grain being eaten by humans? What, besides human consumption, is a major use of grain in the United States? How does this use relate to world hunger? What barriers to food distribution exist, and how can food be distributed more equitably? What alternative food sources exist? How can land be used most efficiently and productively, yet sustainably? What percentage of harvested grain do you think spoils before it is eaten? How might such losses be minimized? (And how can you determine how much food is wasted at your school?)

3. On what chemicals and technologies did the green revolution depend? Why might some of these chemicals be less effective than they once were? How might the availability of water be playing a role? What threats to soil fertility might be affecting yields? Can you think of reasons farmers in developing countries might abandon tractors and "modern" crops and return to their traditional ways of farming?

Data Interpretation:

1. 342 kg per person – 314 kg per person = 28 kg per person;
   28 kg per person/342 kg per person = .082 = 8.2%

2. For India, total beef consumption = 1 kg per person × India's population
   For Italy, total beef consumption = 26 kg per person × Italy's population
   For the U.S., total beef consumption = 45 kg per person × the U.S. population

Chapter 19: Air Pollution

Multiple Choice:

1. c
2. b
3. e
4. a
5. a
6. e
7. e
8. d
9. e
10. b
11. b
12. c
13. a
14. c
15. c
16. d
17. a
18. e
19. d
20. a
21. d
22. b
23. d
24. e
25. e
26. c
27. b
28. b
29. d
30. a

Answers

Matching:

1. d
2. e
3. g
4. c
5. f
6. h
7. a
8. a
9. e
10. f
11. c
12. h
13. b
14. e
15. f
16. d

Fill-In:

1. nitrogen, oxygen
2. Air pollution
3. Primary air pollutants
4. Secondary air pollutants
5. motor vehicles, industries
6. chemical, metals, paper
7. Particulate matter
8. Air toxics
9. smog
10. thermal inversion
11. urban heat island
12. dust domes
13. Clean Air, 1970
14. California
15. respiratory disease
16. global distillation effect
17. sick building syndrome
18. Reading Prong
19. noise
20. decibels
21. hair cells, cochlea
22. bees
23. hydrogen

Short Answer:

1. In Chattanooga, permits were required for open burning, limits were imposed on industrial odors and particulate matter, visible auto emissions were banned, the sulfur content of gasoline was reduced, and sulfur oxide production was controlled. Also, an emissions-free electric bus system was created, and a recycling program was implemented—instead of an incinerator—for dealing with solid waste.

2. The seven major types of air pollutants are (1) particulate matter, (2) nitrogen oxides, (3) sulfur oxides, (4) carbon oxides, (5) hydrocarbons, (6) ozone, and (7) air toxics.

3. Microscopic particles are inhaled more deeply into the lungs, making it more difficult for the body to remove them and increasing the risk of health problems.

4. All three pollutants irritate the respiratory tract, causing airways to constrict. This constriction impairs the lungs' ability to exchange gases and is particularly hard on people with asthma or emphysema.

5. Temperature inversions are most likely in cities located in valleys, near the coast, or on the leeward side of mountains.

6. The EPA established maximum acceptable concentrations for (1) lead, (2) particulate matter, (3) sulfur dioxide, (4) carbon monoxide, (5) nitrogen oxides, and (6) ozone.

7. California, Canada, and the European Union have set a limit of 40 ppm on gasoline's sulfur content. California, Connecticut, New York, and New Jersey require emissions tests for diesel trucks and buses.

8. Developing countries are becoming more industrialized. Environmental quality, however, is normally a low priority; air pollution laws may be ignored or nonexistent. Worsening the situation are the outdated, polluting technologies that are often adopted. Add to these problems rapid population growth, growing automobile use, and the burning of leaded gasoline, coal, wood, and animal dung: the stage is set for increasingly dirty air.

9. One source of particulate pollution is dried fecal matter from millions of gallons of sewage dumped near the city. Another source is wind erosion, which has led to the corrective measure of planting trees on hillsides surrounding the city.

10. The most common indoor air pollutants are radon, cigarette smoke, carbon monoxide, nitrogen dioxide, formaldehyde, household pesticides, cleaning solvents, ozone, and asbestos. (Other pollutants are viruses, bacteria, fungi, dust mites, and pollen.)

Answers

11. If the electricity that powers an electric car comes from a coal-burning power plant, more emissions are released (in this case, from the plant, not the tailpipe) to operate an electric car than a gasoline-powered car.

12. The leading pollution sources are given in parentheses: carbon monoxide (transportation), sulfur oxides (non-vehicular fuel combustion), hydrocarbons (industrial processes other than burning fuel), nitrogen oxides (non-vehicular fuel combustion), and fine particulates (construction/roads).

Critical-Thinking Questions (Hints):

1. How does the topography of the L.A. area contribute to its air pollution problem? How does the population of L.A. compare to that of other large U.S. cities? How about the number of cars and the distance they are driven? What further steps can be taken to reduce car use as well as emissions? Do you think it is feasible to move polluting industries out of the L.A. basin? Is it fair to force companies and individuals to pay for expensive anti-pollution technologies?

2. What does precipitation contain besides water? How does acid deposition affect bodies of water, such as the Chesapeake Bay? What kinds of particulate matter are harmful to water quality?

3. Has your home or dorm been checked for radon and asbestos? Do you or members of your family smoke or use pesticides indoors? What chemicals are used for cleaning? Are safer substitutes available? Do you have a gas range? How often do you open your windows? If you have a window fan, does it blow the outside air in or the inside air out? Does it make any difference? Is air-conditioned and heated air coming from inside or outside the building? How can you find out what fungicides and other chemicals are part of the make-up of your carpets?

Data Interpretation:

1. 18.1 million – 5.4 million = 12.7 million
   18.1 million/5.4 million = 3.35 = 335%

2. 10,000 miles per year per car/25 miles per gallon = 400 gallons per year per car;
   400 gallons per year per car × 100 million cars = 40,000,000,000 gallons per year = 40 billion gallons per year;
   10,000 miles per year per car/80 miles per gallon = 125 gallons per year per car;
   125 gallons per year per car × 100 million cars = 12,500,000,000 gallons per year = 12.5 billion gallons per year;
   amount saved = 40 billion gallons per year – 12.5 billion gallons per year = 27.5 billion gallons per year

Chapter 20: Global Atmospheric Changes

Multiple Choice:

1. a
2. d
3. c
4. e
5. a
6. c
7. b
8. b
9. a
10. b
11. d
12. a
13. e
14. c
15. e
16. b
17. d
18. d
19. d
20. d
21. e
22. e
23. a
24. c
25. c
26. b
27. a
28. e
29. a

Matching:

1. c
2. c
3. c
4. b
5. c
6. c
7. c
8. c

Answers

9. e
10. c
11. f
12. e
13. a
14. d
15. g

Fill-In:

1. agroforestry
2. greenhouse effect, enhanced greenhouse effect
3. models
4. data, assumptions
5. economic growth
6. 18, 7, thermal expansion
7. evaporates, energy, storms
8. 80, zooplankton, global warming
9. numbers, ranges
10. Canada, the Russia Federation
11. 26, 2020
12. population size
13. Carbon management
14. adaptation
15. Ultraviolet (UV)
16. Antarctica, 1985
17. circumpolar vortex, ozone
18. Malignant melanoma
19. acidic, basic
20. hydrogen, positively
21. forest decline, higher
22. are

Short Answer:

1. The IPCC is the U.N. Intergovernmental Panel on Climate Change. It was organized by governments around the world to assess the causes, effects, and extent of global climate change.

2. The following are results of global warming: earlier springs and later falls in the Northern Hemisphere, increased frequency of extreme-heat stress events and droughts, higher sea levels, melting permafrost, retreat of glaciers and polar ice caps, increased frequency of severe storms (including hurricanes and tornadoes) and floods, altered precipitation patterns and agricultural practices, spread of certain tropical

diseases, flooding of coastal areas and low islands, relocation of millions of people, altered ranges of many plants and animals, worsening water shortages in some areas, more frequent El Niños and more intense La Niñas, disrupted oceanic food webs (with declines in many species), extinctions, worsening infestations of certain troublesome species (including weeds and insect pests), more heat-related human illnesses and deaths, decreasing moisture in many agricultural soils, greater increases in nighttime than daytime temperatures, and both declines and increases in regional agricultural productivity.

3. Bangladesh, Egypt, Vietnam, and Mozambique are particularly vulnerable to rising sea level.

4. Polar seas, coral reefs, mountain ecosystems, coastal wetlands, tundra, and boreal and temperate forests are at greatest risk of species loss.

5. Tropical diseases of concern are malaria, dengue fever, yellow fever, Rift Valley fever, cholera, and viral encephalitis.

6. Some critics fear that curtailing carbon-dioxide emissions will hurt the U.S. economy. Others believe the treaty is not strong enough to prevent dramatic climate changes and point out that even if the Protocol is strictly observed, the climate in 2050 will be only 0.1°F lower than it would be without restrictions on carbon-dioxide emissions.

7. These are possible mitigation strategies: slower (or zero) human population growth, increased efficiency of appliances and motor vehicles, carbon taxes and the elimination of energy subsidies, planting trees, carbon management (which involves collecting and sequestering carbon dioxide produced during fossil-fuel consumption), and fertilization of the oceans with iron (thereby stimulating the growth of phytoplankton, which remove carbon dioxide from the atmosphere).

8. Additional compounds that destroy ozone are halons, methyl bromide, methyl chloroform, carbon tetrachloride, and nitrous oxide. The main CFC substitutes are hydrofluorocarbons (HFCs), which are potent greenhouse gases, and hydrochlorofluorocarbons (HCFCs), which attack ozone less severely than CFCs do.

9. Birds rely on calcium in their diets to make strong eggshells and to provide sufficient calcium for the developing bones of their young. This calcium is obtained from insects and snails, which obtain *their* calcium from plants. Plants, however, may not be able to absorb adequate amounts of calcium from soils affected by acid deposition, which causes calcium and other plant nutrients to wash out of the soil. Birds without enough calcium in their diets lay thin-shelled eggs that may break or dry out before the young hatch.

10. Acid deposition, tropospheric ozone, ultraviolet radiation, toxic heavy metals, insect damage, and severe weather may all contribute to forest decline.

11. Precipitation in parts of the Midwest, Northeast, and Mid-Atlantic regions has become less acidic, mostly because of reductions in sulfur emissions. Meanwhile, nitrogen-oxides emissions have risen slightly during the 1990s because of more people driving more cars (many of which are gas guzzlers) longer distances.

Critical-Thinking Questions (Hints):

1. How might altered patterns of precipitation and changes in the rate of evaporation of seawater and the frequency and severity of storms relate to this question? How about temperature change itself, as well as rising sea level?

2. In what ways are developing nations more vulnerable than highly developed nations to the effects of global warming? (See Short-Answer question #2.) What responsibility, if any, do you feel highly developed nations have to help developing nations cope with these effects? What kinds of aid, if any, seem most useful and appropriate?

Data Interpretation:

1. 280 ppm/367 ppm = .763 = 76.3%

2. 130,000 miles/28 miles per gallon = 4642.9 gallons burned; 4642.9 gallons × 5.5 pounds carbon dioxide per gallon = 25,536.0 pounds of carbon dioxide released; 25,536 pounds/2000 pounds per ton = 12.8 tons of carbon dioxide released

3. The pH scale is logarithmic. That means a pH of 3 is 10 times more acidic than a pH of 4, and a pH of 4 is 10 times more acidic than a pH of 5. Multiply 10 × 10 to get the acidity difference between pH 3 and pH 5: 10 × 10 = 100, so a solution with a pH of 3 is 100 times more acidic than a solution with a pH of 5.

Chapter 21: Water and Soil Pollution

Multiple Choice:

1. b
2. b
3. c
4. b
5. e
6. a
7. d

8. e
9. d
10. b
11. d
12. c
13. c
14. a
15. d
16. e
17. e
18. a
19. c
20. c
21. d
22. b
23. c
24. a
25. a
26. b
27. b
28. a
29. e
30. b
31. a

Matching:

1. f
2. g
3. b
4. k
5. n
6. j
7. a
8. h
9. e
10. d
11. l
12. c
13. m
14. i

Fill-In:

1. oxygen
2. biological oxygen demand (BOD)
3. fecal coliform
4. tertiary
5. mercury
6. thermal
7. oligotrophic
8. Artificial (or cultural) eutrophication
9. nitrogen, phosphorus
10. nonpoint, point
11. storm sewers
12. Combined sewer overflow
13. secondary sludge
14. methane
15. National emission limitations
16. hazardous wastes
17. five
18. sewage (untreated)
19. human bodies
20. village well
21. arsenic
22. agricultural
23. soil remediation

Short Answer:

1. Sediment pollution comes from agricultural lands, forest soils exposed by logging, degraded stream banks, overgrazed rangelands, strip mines, and construction sites.

2. The decomposition of sewage by microorganisms requires the consumption of oxygen, which raises the BOD (biological oxygen demand). Therefore, greater rates of decomposition correspond with lower levels of dissolved oxygen.

3. The water supply became contaminated with *Cryptosporidium*, a microorganism that causes diarrhea.

4. Lead comes from lead-based paint (banned in the United States in 1978), volcanoes, leaded gasoline (banned in the United States in 1986), incinerator ash, old factories without pollution-control devices, pesticide and fertilizer residues on food, lead solder on food cans and water pipes, some types of dinnerware, and corroding lead water pipes.

5. Mercury poisoning can harm fetuses in ways that lead to mental retardation, cerebral palsy, and developmental delays. Prolonged exposure causes kidney disorders and severe damage to the nervous and cardiovascular systems; low levels of mercury in the brain cause headaches, depression, and belligerent behavior. People ingest mercury primarily by eating contaminated fish and marine mammals.

6. Pesticides, fertilizers, animal wastes, plant residues, and sediments come from agriculture.

7. Sewage sludge is disposed of through anaerobic digestion, use as fertilizer, incineration, ocean dumping, and disposal in landfills.

8. Many small, easily overlooked sources of pollution have a large combined effect. Also, government agencies do not have enough money or staff to enforce environmental laws.

9. The Refuse Act (1899), the Clean Water Act (first passed in 1972 as the Water Pollution Control Act), the Safe Drinking Water Act (1974), the Resource, Conservation, and Recovery Act (1976), and the Ocean Dumping Ban Act (1988) protect water quality in the United States.

10. Signs of severe pollution included eutrophication, high bacterial counts, fish kills, and severe birth defects in many animal species.

11. The Great Lakes are "contaminated" with exotic species, including zebra mussels and sea lampreys, which threaten native species. Ongoing shoreline development increases sediment pollution by contributing to flooding and erosion. Furthermore, many toxic substances, including mercury, remain in the lakes.

Critical-Thinking Questions (Hints):

1. Think about pressure put on governments to clean up rivers or to provide public funds for wells, water treatment facilities, etc. What conflicting demands might governments face? Who owns the water in a river that flows through two or more countries? What economic burdens result from a shortage of safe drinking water? Who bears them? What emotional costs are borne and by whom?

2. What kinds of businesses and industries can you think of that profit by polluting or lose money by not polluting? How can governments keep a balance between safeguarding natural resources and human health on the one hand, and safeguarding individual rights on the other? What are the economic benefits of having clean water and air? What other benefits are there? Who should pay for enforcement of environmental laws? How do *you* feel about the fairness, necessity, and effectiveness of environmental laws?

3. See "You Can Make a Difference: Preventing Water Pollution" in your text. What motivates and instructs us to buy certain products? Why might we think an inexpensive product, such as baking soda, is less effective than a more expensive product? Why do you think we don't see TV commercials about the cleaning power of baking powder or vinegar?

Data Interpretation:

1. $25 million – $7 million = $18 million

2. Number of seriously contaminated watersheds = $1363 \times 7\% = 1363 \times .07 = 95$

3. $0.0141/0.0065 = 2.17$ times more oxygen at zero degrees (Celsius)

Chapter 22: The Pesticide Dilemma

Multiple Choice:

1. d
2. c
3. c
4. a
5. e
6. e
7. b
8. d
9. b
10. a
11. a
12. c
13. c
14. c
15. e
16. e
17. e
18. d
19. a
20. b
21. a
22. b
23. d
24. a

25. e
26. c
27. d
28. d
29. b
30. e

Matching:

1. d
2. g
3. b
4. l
5. m
6. a
7. h
8. i
9. e
10. c
11. n
12. j
13. k
14. f

Fill-In:

1. narrow-spectrum, broad-spectrum
2. Nonselective, selective
3. dioxin
4. pathogens
5. genetic resistance
6. pesticide treadmill
7. Biological magnification (or biological amplification), bioaccumulation (or bioconcentration)
8. *Our Stolen Future*
9. Japanese beetle
10. 10, 20
11. genetically modified (GM)
12. 20
13. quarantine
14. Integrated pest management (IPM)
15. pesticide, rice
16. Pesticide Chemicals, Miller
17. Federal Insecticide, Fungicide, and Rodenticide

18. groundwater, inert
19. Food Quality Protection
20. Food and Drug, 1
21. pesticides
22. Calendar spraying, scout-and-spray

Short Answer:

1. "Perfect" pesticides would be narrow-spectrum (would not harm species other than the one targeted as a pest), would be readily degraded into safe materials, and would not move around in the environment (including food webs).

2. *chlorinated hydrocarbons*: DDT, endosulfane, lindane, methoxychlor
   organophosphates: methamidophos, dimethoate, malathion
   carbamates: carbaryl, aldicarb

3. A variety of soft-tissue cancers, skin diseases, urological disorders, and birth defects are linked to dioxin exposure.

4. Pesticides can kill pest predators directly; they can kill them also by reducing their food supply (the pests) and through the poisoning effect of eating pesticide-contaminated pests. In addition, pest predators in search of food may leave a sprayed area for more successful hunting elsewhere. The reduction of pest predator populations is one reason heavy pesticide use has not significantly reduced crop losses to pests. Other reasons include increasing genetic resistance to pesticides and changes in agricultural practices, such as reduced use of crop rotation.

5. Often, dangerous pesticides that have been banned in highly developed nations are shipped to developing nations. Also, many farmers in developing nations are not trained to handle and store pesticides safely, and safety regulations are typically lax. Unwanted stockpiles of pesticides are accumulating, too, especially in developing nations; chemicals leach from storage drums and contaminate waterways and groundwater.

6. Children tend to play on floors and lawns where they are exposed to pesticide residues. Also, children's bodies are still developing, which may make them more susceptible to pesticide poisoning. Brain cancer and leukemia are two of the suspected health risks children face.

7. A weevil introduced in 1968 to control the Eurasian musk thistle feeds also on a threatened North American thistle. The sterile male technique is expensive and must be carried out continually to prevent resurgence of the pest population.

8. Some insects appear to be developing resistance to the *Bt* toxin.

9. Pheromones can be used to lure insects, such as Japanese beetles, into traps, where they are killed. Or, pheromones can be released into the air to confuse insects, preventing them from finding mates.

10. IPM practices are still being worked out for weed (as opposed to insect) control and for crops in specific locations with unique climate and soil conditions. Also, IPM requires a great deal of knowledge, which many farmers have not yet gained.

11. Aldrin and dieldrin (1974), kepone (1977), and chlordane and heptachlor (1988) have been banned by the EPA.

12. Consumers typically demand perfect, unblemished produce, which encourages farmers to use pesticides. Unless consumers begin to demand pesticide-free produce, farmers will continue to provide what sells: blemish-free produce grown with the aid of pesticides.

13. Substitute pesticides may not exist or may not be as effective. They may also be more expensive than traditional pesticides. There are also economic concerns: a pesticide ban would increase food prices and impose economic hardship on some farmers.

Critical-Thinking Questions (Hints):

1. Why would a species (such as humans) with a long generation time evolve more slowly than a species (such as a type of insect) with a short generation time? What is the advantage of a large population, compared with a small population, in terms of evolutionary change?

2. What effect might a pesticide ban have on crop yields? On crop spoilage? How could some farmers benefit, at the expense of other farmers and food prices, if a ban were imposed?

Data Interpretation:

1. $2,700,000/6,000,000,000 = 0.00045 = 0.045\%$

2. (See Table 22-2 in *Environment*, 3/e.)
   1974 percentages: $13 + 12 + 8 = 33\%$
   1989 percentages: $13 + 12 + 12 = 37\%$

Chapter 23: Solid and Hazardous Wastes

Multiple Choice:

1. c
2. d
3. e
4. a
5. a
6. b
7. a
8. e
9. d
10. c
11. c
12. a
13. e
14. b
15. c
16. d
17. d
18. b
19. e
20. e
21. e
22. b
23. c
24. a
25. d
26. a
27. c
28. c
29. a
30. c

Matching:

1. c
2. j
3. e
4. i
5. a
6. f
7. g

8. h
9. b
10. i
11. d
12. h
13. h

Fill-In:

1. Municipal solid waste
2. Sanitary landfills
3. paper products, yard waste
4. compacted clay, plastic sheets
5. methane, electricity
6. plastic
7. paper, plastics, rubber
8. dioxins
9. Modular, mass burn, refuse-derived fuel
10. Pollution Prevention
11. dematerialization
12. durable
13. Denmark
14. economics
15. fee-per-bag
16. dioxins
17. PCBs
18. inherent safety
19. Superfund
20. lawyers' (legal)
21. Environmental (or Green)
22. Environmental justice
23. Integrated waste management
24. voluntary simplicity
25. product stewardship

Short Answer:

1. Municipal solid waste consists primarily of paper and paperboard, yard waste, plastics, metals, wood, food waste, glass, rubber, leather, and textiles.

2. Among the factors that go into landfill-site selection are geology, soil drainage, proximity of wetlands and bodies of water, distance from population centers, precipitation, and likelihood of flooding.

3. Thousands of landfills have been closed because they reached their capacity or failed to meet environmental standards. The construction of new landfills is often blocked by local citizens; also many "desirable" sites have already been claimed, thereby limiting location choices.

4. Tires tend to move upward to the surface of a landfill. There, they collect rainwater and serve as mosquito breeding "ponds." Tires also pose a fire hazard.

5. The three goals of waste prevention are to (1) reduce the amount of waste, (2) reuse products, and (3) recycle materials.

6. Refillable bottles are thicker than one-use bottles; their increased weight increases the cost of transporting them. Also, there is added expense associated with sterilizing used bottles.

7. The Toronto city council passed a law requiring that daily newspapers—in order to use city vending boxes—contain at least 50% recycled fiber; President Clinton issued an executive order that all paper purchased by the federal government must have 30% recycled content.

8. Recycled tires can be used for playground equipment, trashcans, garden hoses, coastal reef-building structures, and rubberized asphalt.

9. Hazardous chemicals include acids, dioxins, abandoned explosives, heavy metals, infectious waste, nerve gas, organic solvents, PCBs, pesticides, and radioactive substances. Texas, Louisiana, and California had the most toxic-chemical accidents in 1998, numbering 1392, 644, and 215, respectively.

10. Lead and other heavy metals, trinitrotoluene (TNT), selenium, certain pesticides, and radioactive strontium and uranium can be removed or degraded by plants.

11. The most effective way to manage hazardous waste is by source reduction; next-most-effective is conversion to less hazardous materials (toxicity reduction); next is long-term storage.

Critical-Thinking Questions (Hints):

1. What costs are associated with delivering trash to landfills? Recycling glass? Cleaning up parks, schoolyards, and roadsides? What other costs must society bear when beverages come in one-use bottles? Who bears these costs?

2. How might the export of used paper encourage or discourage recycling in the United States? How might it affect domestic markets for recycled paper? What impact might the export of used paper have on the timber industry and the fate of forests in the United States?

3. What kinds of federal and state programs are likely to be cut when recession occurs? How can recession cause people's priorities and behaviors to change in ways that affect recycling? (For example, how does the cost of recycled notebook paper or paper towels compare to the cost of non-recycled items?)

Data Interpretation:

1. If 1 ton of recycled paper saves 17 trees and 7000 gallons of water, 30 tons of recycled paper saves $17 \times 30 = 510$ trees and $7000 \times 30 = 210,000$ gallons of water.

2. Use the *World Population Data Sheet* to find the U.S. population. Divide 65 million by this number. Multiply the quotient by 100 to obtain percent.

Chapter 24: Tomorrow's World

Multiple Choice:

1. d
2. e
3. b
4. a
5. a

Fill-In:

1. 4.5, 25
2. four, eight
3. 2.5, 6
4. sustainable
5. biological diversity
6. development, resources
7. 10, 80
8. Conservation tillage
9. wastes
10. consumption, waste
11. sunlight, heat
12. cities
13. Earth Summit, international
14. problems, opportunities
15. attitudes, values
16. cost

Short Answer:

1. These degraded resources are productive living systems, such as forests, shrublands, grazing lands, and wetlands.

2. We rely on such ecosystem services as protection of watersheds and topsoils, development of fertile agricultural lands, determination of local and global climate, and habitats for the animals and plants we rely on.

3. In many parts of the world, women are expected to bear the whole responsibility for child care and, at the same time, make a substantial contribution to family income through direct labor. They may also have few legal rights to their income, their property, or their children.

4. Negative effects of agriculture include soil and water pollution, air pollution, soil erosion, loss of soil fertility, and aquifer depletion.

5. Among the citizens of developed nations, there exists apathy and ignorance about the state of the environment. Many people feel detached from and "above" the natural world, as if its fate and ours are not connected. Also, there are strong incentives (provided, especially, by the media) for wasteful consumption. In developing nations, many people are too concerned about their immediate survival to be able to make personal sacrifices or efforts for the environment. Too often, throughout the world, people do not believe that their individual actions can make a real difference; they just give up.

6. We can establish environmental curricula in schools and colleges; support the activities of environmental organizations; support natural history museums, zoos, aquaria, and gardens that promote conservation and sustainability; and encourage environmental activism in church programs, social groups, and other settings.

7. Brant referred to the flow of money from developing nations to highly developed nations, which has exceeded the flow in the other direction for more than 15 years.

8. We (in highly developed nations) deplete resources throughout the world. In other words, our lifestyles are not based on the concept of sustainability; we live off Earth's dwindling principle, not interest.

Critical-Thinking Questions (Hints):

1. Why do people today, on average, spend less time outdoors than their ancestors did? Do you agree or disagree that one must "know" something to love it? How are we insulated and isolated from environmental damage and the costs it entails? Have you ever seen a clear-cut forest? A badly polluted river? A Superfund site? An endangered species?

2. Today, people are concentrated in cities, largely because that's where most of the jobs are. How might telecommuting change that picture? Can it contribute to urban sprawl and the loss of farmland and wildlife habitat?

Data Interpretation:

1. $x/.25 = 1.8$ billion tons$/.15$; therefore, $x = (1.8$ billion tons$/.15)(.25) = (12$ billion tons$)(.25) = 3$ billion tons